Genetic Heritage

by Ken McLean

Leading sire in the world by earnings for 1995 was Breeder's Cup Classic Gr-1 winner SUNDAY SILENCE (by Halo) which stands at stud in Japan. He has an affinity with mares carrying the strain of Northern Dancer.

Other thoroughbred breeding books written by the author:

"Quest For a Classic Winner"

"Tesio - Master of Matings"

MR. PROSPECTOR (Raise a Native-Gold Digger by Nashua) carries topweight to victory in one of his stakes victories. Leading sire in North America, he sired Woodman, Seeking the Gold, Miswaki, Gone West, Machiavellian and Gulch.

"GENETIC HERITAGE"

First Edition

Published by: K & C McLean,
Ken McLean Bloodstock Ltd
P. O. Box 8558
Lexington Kentucky 40533
U.S.A.

ISBN 0-9619432-1-1

Printed by: Accuprint, Inc.
2414 Palumbo Drive
Lexington Kentucky 40509
U.S.A.

Phone: (606) 268 8844

Fax: (606) 266 0317

European champion filly and U.S. champion miler MIESQUE (Nureyev-Pasadoble) bred and raced by the late Stavros Niarchos and now under the care of his daughter Maria. Miesque has already produced French 2,000 Guineas Gr-1 winner Kingmambo.

"Every racehorse mating is affected by genetic probability factors. The hybrid thoroughbred breed possesses wide genetic variation and thousands of recessive or hidden characteristics wait patiently to be exposed in an individual's genotype with the right combination of parents. Pedigree research makes one aware of incredible odds needed for any person to be successful at breeding a champion racehorse or major stakes winner.

By manipulating the genes of a newborn foal, through duplication of superior ancestors, racehorse breeders can increase the odds for successful matings since this technique challenges Mother Nature to produce upgraded performance. Famous men and women throughout thoroughbred history who cleverly achieved the task of breeding a champion more than once usually made their own luck via the process of careful selection.

They were animal breeders extraordinaire."

- Ken McLean

Contents

Page

Foreword ***1***

Acknowledgments ***3***

Introduction ***4***

Chapter 1
The Thoroughbred Gene Pool - Origins ***8***

Chapter 2
Great ancestors which shaped the modern racehorse ***38***

Chapter 3
Speed and Stamina factors... ***60***

Chapter 4
Hereditability of heart size and the "X" factor.................. ***71***

Chapter 5
Male line survival - Fertility and its cyclic function.......... ***87***

Chapter 6
The Mitochondria of cells and its maternal importance.. ***114***

Chapter 7
Blandford and his influence for stamina ***127***

Chapter 8
Purchasing foundation breeding material - buying from public auctions... ***135***

Chapter 9
Successful examples of Inbreeding................................ ***141***

Europe's horse- of- the-year NASHWAN (Blushing Groom-Height of Fashion by Bustino) winner of the English 2,000 Guineas Gr-1 and Derby Gr-1.

Chapter 10
Planning matings for the racetrack *156*

Chapter 11
Myths and misconceptions *160*

Chapter 12
Blushing Groom's family.. *168*

Chapter 13
Famous genetic affinities called "nicks"........................... *174*

Chapter 14
The Maktoum and Niarchos empires.................................. *184*

Chapter 15
Significant Breeders of the 20th Century........................... *194*

Chapter 16
Asian markets and their impact in the
21st Century.. *206*

Chapter 17
Champions of the Century.. *208*

Bibliography.. *217*

Index of Pedigrees... *218*

General Index... *219-230*

Northern Dancer's champion son NUREYEV. Winner of the English 2,000 Guineas Gr-1, he was susequently disqualified for interference. He is the sire of Theatrical, Miesque, Soviet Star, Zilzal, Robin des Pins, Sonic Lady, Mehthaaf, Gayane, Wolfhound and Polar Falcon. He is half-brother to the dam of England's leading sire Sadler's Wells.

Foreword

This third book written by the author covers the establishment of the thoroughbred breed, listing the most important Sires and Dams which shaped the modern racehorse.

Ken has done a considerable amount of research which no doubt will be useful information for younger people coming into our industry. His previous publication "Quest For a Classic Winner" was so popular it was necessary to reprint it.

He has always been fascinated with the genealogy of the racehorse, and this book reflects his passion for the subject.

- Sir Philip Payne-Gallwey Bt

Sheikh Maktoum bin Rashid al-Maktoum, the king of Dubai leading in CARNEGIE, after the colt's victory in Longchamp's valuable Prix de l'Arc de Triomphe Gr-1.

Acknowledgments

The author wishes to express his deep gratitude to everyone who helped shape his career with thoroughbred breeding and racing. In particular, I wish to thank Josephine Abercrombie and Pin Oak Farm's manager Clifford Barry, Maria Niarchos and Fresnay-Le-Buffard's manager Tim Richardson, my two mates at the British Bloodstock Agency namely Sir Philip Payne-Gallwey and Joss Collins, as well as New Jersey friends Leonard, Lois and Jon Green.

I owe a debt of gratitude to John & Susannah Prather and Monty Hinton, my Kentucky friends who first introduced me to one of the kindest men in the business, William T. Young, owner of Overbrook Farm.

Many European, Australian and North American friends have helped the author gain valuable experience and special thanks go to Mr. and Mrs. William T. Young, John and Beth Phillips, David and Betsy Hager, Robert and Blythe Clay,Mr. and Mrs. Arthur B. Hancock III, Brereton Jones, Jack Kent Cooke, John and Alice Chandler, Robert Berger, Michael Goodbody, John Leaver, John Messara, Dr. Richard Holder, His Excellency Nasser Abdulla Hussain Lootah, Antony Beck, Duncan Taylor, Bobby Spalding, Brian Cross, Mike Whitehouse, The Blood-Horse Magazine, The Thoroughbred Times, Andrew Reichard, and our dear friends Peggy and Les Bridge.

To my loving wife Carla and our daughters Kiersten and Tanya who encouraged me to write a third book, supplied ideas, compiled the index, many thanks from the deepest part of my heart.

Finally, I wish to thank Lady Wentworth, author of "Thoroughbred Racing Stock" whose persistent and painstaking research helped clarify the origins of foundation breeding stock listed and unlisted in the initial General Stud Book.

It was Sir Philip Payne-Gallwey and Sydney trainer Les Bridge who encouraged me to really think about the methods of breeding and training the racehorse, and so I dedicate this book to these two gentlemen.

INTRODUCTION

" Oh for a Horse with wings." - William Shakespeare

Breeding the racehorse is an inexact science and shall remain so until geneticists can supply accurate information about individual genetic codes, i.e. gene maps. The transmission of inheritance is one of the wonders of the world and because the modern racehorse is a hybrid animal, it rarely breeds true-to-type.

The laws of heredity were discovered by the Austrian monk, Gregory Mendel, and published in 1866. His contribution to the science of genetics was profound and his experiments enabled him to formulate principles which lasted the test of time. Using garden peas he selected 34 varieties and chose from them a number which showed distinct, contrasting forms. He was the first person known to record the numerical results and discover certain consistencies - hereditary factors.

Mendel discovered the difference between "dominant" and "recessive" genes. A gene is the fundamental unit of inheritance and he noticed that when the pollen of a smooth seed and a wrinkled seed were exchanged between plants, regardless of which plant bears them, all the seeds resulting from this cross were smooth. Repeated experiments confirmed this consistent result and Mendel determined that the smooth seed was always *dominant* over the wrinkled seed. However, seeds resulting from the cross of smooth and wrinkled seeds are not pure-breds or true breds. Their genetic composition is different. Although in appearance the seeds are smooth like the original pure-bred parent, they in fact contain the hidden characteristic for wrinkled shape seeds. The wrinkled shape seed is *recessive*, waiting to be exposed in future generations.

To simplify this example, let's give a code for the parents of the seeds. The smooth seed may be described as SS and the wrinkled seed ww. Capital letters are used for a dominant trait and lower case letters for a recessive trait. The experiment showed that when SS is crossed with ww, 100% of the seeds produced are Sw (all smooth, but masking the trait for wrinkled seed). When these first generation seeds Sw are cross-pollinated, they produce the hidden wrinkled shape recessive trait in a certain proportion or ratio. This occurs because of the possible combinations:

First Cross: SS x ww = 100% Sw
(smooth) (wrinkled) (smooth in appearance, masking wrinkled)

Second Cross: Sw x Sw = 25% SS, 25% Sw, 25% Sw and 25% ww

or 25% SS, 50% Sw, 25% ww
(smooth) (smooth/w) (wrinkled)

ratio: 1 : 2 : 1

The parent SS receives its S genes from its own parents; and similarly, the parent ww receives its w genes from its own parents. Each character is controlled by a gene. Genes exist in alternative forms called *alleles* which control the alternative traits of a character. The *alleles* used in the

second cross are S and w in both parents, and when mated together, combine to form three results namely S+S, S+w, and w+w. You may wish to substitute SS and ww for the coat color of the racehorse, for example, the color brown is *dominant* and the color chestnut is *recessive*. When BB (dominant brown alleles) are matched with cc (recessive chestnut alleles) it will produce 100% Bc (brown foals in appearance, but always masking the chestnut coat color). When two brown thoroughbreds with alleles Bc are matched together four times, normal expectant results will deliver one foal which is pure brown or BB, two which are brown in appearance but impure Bc that mask the chestnut trait, and one foal which is chestnut or cc (inheriting both recessive alleles).

Imagine the *thousands* of different combinations which occur at conception between two thoroughbreds with wide genetic variation. There are countless combinations of *dominant and recessive* traits involved in the formation of a foal's genotype or genetic make-up. In the opening Chapter, I briefly explain the formation of the so-called "Thoroughbred" horse because the reader should be made aware that although the world's keepers of the Stud Book "closed" the gene pool in the 19th Century, genetic variation has widened because of a racehorse population explosion which occurred after 1945. The racehorse became a popular breed on all continents of the globe and its evolution has taken different routes throughout the 20th Century. Generally speaking, the genetic make-up of racehorses born in Germany up to 1980 are uniquely different to successful thoroughbred strains found in other countries such as Australia, New Zealand, South Africa and Argentina.

Successful male lines flourished in new regions because they were able to adapt to a new environment, climate, altitude, feeding program etc. whereas the same lines became almost extinct in their birthplace of origin. Successful female lines sprang up in distant thoroughbred breeding countries from taproots that never could flourish in either Europe or North America. The *internationalism* of the racehorse eventually caused the world's best strains to become intermingled. Despite an initially restricted gene pool via European and North American stud books, the racehorse as a breed possesses wide genetic variation, mainly as a result of recessive genetic expression. Providing artificial insemination is never introduced, I envisage the racehorse of the 21st Century will maintain a healthy outlook, thereby sustaining wide genotypic variation for the future of the breed. I recall a statement full of commonsense, made by Walter D. Osborne, and I quote:

"The Thoroughbred horse is unquestionably man's finest achievement in the art of selective breeding. So finely has this strain of animals been developed over the years that today a mere tick or two of the timer's watch separates the great horse from the run-of-the-mill."

Breeding stakes winning horses is very much dependent upon genetic probability factors. Full brothers and sisters take *different portions* of their parent's genetic make-up and not only inherit unlike phenotypes (visual appearance) but also inherit *variance* in performance on the racetrack. Perhaps the art of successful racehorse breeding is dependent upon the ability of an animal breeder to match a stallion with a mare whereby each parent can upgrade the biomechanics of the other to produce a superior athlete.

For this to happen, however, there must be *compatibility* between both parents - between the stallion and the mare. If *superior* ancestors can be found on both sides of a foal's pedigree there is a high chance genetic *compatibility* will exist. Simply breeding the best to the best has proven to produce foals with differing levels of performance; however, when the best are biochemically incompatible, failure will result. Hidden recessive genes have a habit of surfacing when least expected unless temporarily dominated or masked.

I am fascinated by the haphazard transmission of speed from generation to generation; and am intrigued as to how a Group 1 or Grade 1 racehorse inherits exceptional acceleration and a favorable clearance of lactic acid in its system. Ancestors able to pass superior speed and soundness on to their descendants are indeed *Genetic Giants* of the breed, and hence, their pedigrees should be studied in an attempt to understand their intrinsic genetic make-up.

Amazing though it may seem, not until specific gene mapping is conducted on a large scale can we hope to begin to comprehend how superior racing class is handed down from elite ancestors which represent a tiny portion of the total gene pool. Elite ancestors possess highly superior muscle fibers, large hearts, clean air-flow and strong nervous systems which are able to cope with stress. Even with new scientific gene mapping, will we ever be able to identify unraced thoroughbreds possessing courage and the indefinable *will-to-win?*

"Genetic Heritage" is a breeder's book which merely expresses subjective opinion. It is written in good faith using knowledge and experience gained during 43 years of study and observation and is an attempt to help tomorrow's breeder of the thoroughbred. Real improvement of racehorse speed slowed down in the final half of the 20th Century judging by the rate at which new track records were accomplished. Next century we can expect a continuation of genetic variation and its often surprising results, but our grandchildren might have the benefit of meaningful equine genetic research.

Exciting new information about the transmission of inheritance should be used by skilled animal breeders, acting on advice from veterinarians and genetic consultants. The selection process may require commercial thoroughbred farms to hire specialists in this area to service client needs and optimize results. Needless to say, the magic of breeding the racehorse might never be reduced to computerized spreadsheets of valuable genetic data. After discussions with eminent geneticists, I formed an opinion a few years ago that it is unlikely genetic research will ever be able to forecast a thoroughbred's courage and will-to-win. Not even heart scores can measure these qualities.

Chapter One of this book contains information about the origins of the British thoroughbred. I felt it compulsory for breeders to try to comprehend how the thoroughbred breed was established and how it evolved, because this information may help explain existing diversity of phenotypes. Chapter 6 explains the significance of the Mitochondria of cells, and scientifically announces the importance of direct female transmission - something quite new.

The racehorse is genetically a hybrid animal, and as a consequence, breeding stock does not as a rule breed true-to-type. Therefore, breeding for speed plus stamina presents an enormous challenge. Although success will be governed by the whims of genetic probability factors (chance), I have proven to many clients that one can reduce risk of failure by employing selection skills. I trust there will be an element of sound advice in the pages which follow and that Chapter One is not skipped over because of its historical detail. The origins of the racehorse trace back to the selection skills employed by Bedouin tribesmen, Phoenicians, Sabaeans, Egyptians, Syrians and Persians at least as far back as 3,000 B.C. - probably much longer. These pioneers maintained the purity of the Arabian horse which today represents maybe 75% or more of the entire gene pool of the modern thoroughbred. The racehorse certainly has a unique heritage.

- *Ken McLean, July 1996.*

Unbeaten at three, SEA-BIRD II, grandson of Native Dancer, wins the 1965 English Derby Gr-1 from Meadow Court and I Say. He went on to win the Prix de l'Arc de Triomphe Gr-1 and is the sire of Arctic Tern, sire of French champion Bering.

THE THOROUGHBRED GENE POOL - ORIGINS

"I have called this principle, by which each slight variation if useful, is preserved, Natural Selection." - Charles Darwin

The evolution of the racehorse is a process we need to comprehend. The "Thoroughbred" horse is a poor definition because the racehorse is not a pure-bred and derives its genotype from a variety of different horse breeds. We are told that the size of *Eohippus* approximated that of a fox and supposedly existed about fifty million years ago. *Eohippus* is regarded as the starting point in the evolution of all breeds of horse.

When the Chaldeans, Sabaeans, Phoenicians, Syrians and Bedouins tamed the horse, they realized this animal was of extreme value. Many theories have been put forth as to the original home of the purebred Arabian horse and its descent. I doubt its origin was restricted solely to the land of Arabia, but it was perfected as a breed in Arabia (Arabah, which lay between the mountains and the Red Sea). The descent of the purebred Arabian horse took a different highway to that of other horse breeds including the Tarpan, the Russian Steppe, and the Eurasian native horse. The horse was a hunter, a means of transport, a work animal capable of pulling heavy weights, a swift steed in battle, a crusader, a carnival entertainer, a pleasure horse, and dare I say oftentimes a source of food for man's survival.

When Alexander the Great met the forces of Darius III at Arbela in 331 B.C. this historic battle established the superiority of infantry and cavalry over infantry and chariots. The invading Persians were driven from the Grecian peninsula. By 776 B.C. the Greeks first held the Olympic Games in honor of their god Zeus, establishing a Hippodrome or horse-course. By the 25th Olympiad, chariot horse racing proved to be the biggest drawcard. About 700 B.C. the breeding of fast endurance horses was given priority status among generals and rulers throughout southern Europe, Asia and the Middle East. The Persians adored their fleet horses and Persia's royal studs kept accurate records of pure breeding lines which would be called "Arabian" (*Equus Arabicus*). It is hard to imagine that the fertile lands throughout the middle east 2,000 years B.C. would gradually become desert wastelands as a result of environmental abuse, severe earthquakes, and the ravages of warfare. The once green pastures of Damascus and Aleppo would force the purebred Arabian to be sold to regions of Turkey, Egypt, Greece, Italy, Hungary, and Poland.

From the conquests of war, the Pharaohs of mighty Egypt inherited huge numbers of Asian impure native horses (about 12 hands high) which possessed tough constitutions. Large numbers of Egypt's native horses which survived African horse-sickness were mated with other breeds of horse including purebred Arabian stock, and the size of the Egyptian horse increased above 13 hands. The Assyrian horse was a heavier type, favored by armored crusaders, but it lacked speed. The origin of the Andalusian breed might trace to Assyrian descent. Probably the Greeks and Romans were first to develop the saddle, bridle, spurs and riding boots. The Persians captured purebred Arabians from Bedouin tribes and were responsible in developing the whip, rope halter and the first to institute purebred Arabian stud records. Perhaps the saddle, which was known in existence in Athens and Sparta around 340 B.C., was an adaptation of the Asian saddle devised by Chinese and Mongolian horsemen, or else it was an adaptation of the cloth used by Bedouins. Greek and Roman horses were groomed daily, oil being applied to their coats. Nomadic Bedouins bred many fine horses which probably descended from the wild horses of Tartary.

The Arabian purebred horse came in three distinct phenotypes, namely *Kocklani, Kehidischi* and *Attichi.* The breed known as *Kocklani* (or *Kekjlani)* were kept pure and certificates of matings were witnessed to ensure proof of bloodline. These horses were extremely valuable and mostly in the possession of wealthy merchants, generals, princes and kings. To my knowledge from careful research, there existed at least 16 specific strains of purebred Arabian horses in the Middle East, all developed by Bedouins.

English author Margaret Greely aptly states in her interesting book "Arabian Exodus", the "horse in the wilderness - this is the major chord in all literature and art composed and inspired by the Arabian horse. He is part of that strange fascination which the desert holds for men, and men who sought the grandeur of the desert found there too the horse of the desert".

Other horse breeds developing simultaneously with the Arabian breed include the *Tartar, Russian Steppe horse, Eastern native horse, Spanish riding horse, European native horse, North American native horse, Central European heavy horse,* and the *miniatures from Siberia.* These breeds possessed considerable variation of size, strength, shape, constitution and temperament. Miniature horses were imported from Siberia and China into central Europe to establish numerous pony breeds.

During the 12th Century, the purebred Arabian breed was considered to be the prized riding horse for battle and sport, and its numbers increased. By the 13th Century the whole Arabian Peninsula and the adjoining countries of Syria, Iraq, Lebanon and North Africa lay under Turkish domination, with Egypt as a major command post. Arabian horses imported and bred in Turkey became known as *Turks*, but in more north-east regions they were called *Barbs.* Because purebred Arabian horses commanded high prices, it was common practice not to mate mares with any other breed, and so most of the *Turk* and *Barb* strains retained their Arabian ancestry unspoiled. Gradually there became some distinction between the phenotype (outward appearance) of the *Arabian, Turk and Barb.* Treasured most of all were the purebred Arabian horses of Nejd which could be purchased by rich merchants. Few Arabian stallions or mares were ever made available for sale by the ruling families of Persia and Syria. It was not uncommon for royal families to own up to six thousand purebred Arabian horses. Indeed, there has been an historical record of eight thousand horses owned by the Arab tribe of Montefek pastured on the banks of the Euphrates river, near Nejd.

The Bedouins believed it was improper to mate an Arabian mare until she was five years old because they believed it would draw too much upon her calcium levels at an earlier age, but poorer desert dwellers bred from four year old mares.

Due to custom, the Bedouins never allowed a foal at the moment of birth to touch the ground; instead they received it in their arms, caressing it with gentle strokes to stimulate blood circulation. The foal was washed before it was placed on the ground. The highest virtue of the Arabian horse is its endurance, strength and wind.

Barbs became slightly taller than *Arabians,* and featured a lean head, higher withers, longer pasterns and slight coarseness. *Turks* were usually about 14 hands high, sometimes roman-nosed with longer rein, a more upright shoulder angle, but with longer forearms and gaskins than the *Barb,* but not lacking in refinement.

When the Romans conquered Britain in 56 B.C. they brought with them cavalry horses as well as many very heavy horses from Prussia, Denmark and Belgium. Once established, Roman generals and governors imported a variety of horses from Gaul, Italy, Spain, and the Middle East. The mixture most favored was to match Arabian stallions with mares of either *Barb or Turk* strains.

The Roman governor Severus is supposed to have raced purebred *Arabian* horses in Yorkshire in the year 211 A.D. By the year 900 A.D., German and French horses were imported to England and Scotland so by the time King Richard I prepared for his crusades, Britain's horse population began to flourish.

It was King John who became interested in horse sports and added horse racing to his procession of tournaments. Not until the reign of Queen Anne was horse racing of any significance conducted. Henry VII and Henry VIII promoted foxhunting and *the wild goose chase.* At Smithfield, London, in about 1174, a racetrack was used on a regular basis. By 1512, trophies were awarded the winners of flat racing at Chester.

King James 1st instituted the recording of pedigrees and a handicapping system. In 1619 the first match races took place at Newmarket. King Charles II established the King's Plate in 1665. On August 11th, 1711, Ascot race course was officially opened. Of special note, the mare Bonny Black (founder of the Number 20 Family) won the Ascot Gold Cup defeating all the best older males. During the reign of Queen Anne and William III, the importation of three stallions was to have dynamic affect on the subsequent establishment of a hot-blooded horse now called the "Thorough-bred".

I refer to *The Byerley Turk, The Darley Arabian,* and *The Godolphin Arabian* from which all racehorse sire lines are supposed to descend. These three stallions provided the "Y" male sex chromosome. The Cullen Arabian, imported to England in 1745, became an important broodmare sire. When these important stallions were brought to Great Britain, they were matched with mares sired by at least 174 non-related stallions. In 1780 the first English Derby was run. Sons and grandsons of the three great stallions were exported to Europe and North America before this date. In 1793 the English General Stud Book came into existence. Unfortunately the Thorough-bred was referred to as a purebred, and the trotter mistakenly referred to as a pure standardbred.

No other breed of horse possessed the quality of bone, tendon, muscle, heart and speed of the Thorough-bred. This animal had symmetry, thin skin, high spirit and courage. The blending of many different types of horse created a hybrid racehorse with speed and endurance. Modern racehorses exhibit characteristics of their forebears. The dish-faced racehorse inherits more traits from his *Arabian* ancestors. Examples of this phenotype may be defined as Mahmoud, Mumtaz Mahal, Holy Bull and Palace Music. The plain roman-nosed racehorse inherits more traits from its *Turk* ancestors, and the taller but more muscular short-nosed racehorse owes its type from its *Barb* ancestors. Racehorses with an extra rib tend to have a high proportion of genetic influence from other sources. English breeders during the 18th Century considered imported *Barb* stallions as ideal mates for mares of *Arabian* phenotype because they transmitted high withers.

The famous stallion Eclipse stood 16.2 hands high and helped develop size in the British racehorse. Having researched his pedigree, he possessed much Arabian influence but had a much higher share of Barb ancestors than would normally be found in pedigrees of his contemporaries.

It was not uncommon for races to be run in heats of up to four miles distance and shorter races were established in 1776. Colonel St. Leger proposed an event over 2 miles for three year old colts and the English St. Leger was first run in 1778. By 1887 it was run over 1 mile 6 furlongs and 132 yards. It is the oldest stakes race in the world. The English three year old classics soon followed and became known as The Derby, The Oaks, followed later by mile classics known as the 2,000 Guineas and 1,000 Guineas. In 1875, the Dewhurst Stakes for two year olds over 7 furlongs was instituted at Newmarket.

The first organized racing in North America took place in 1665, and the first official record of racing was held in 1668. Virginia and the Carolinas became the initial breeding nurseries of the Thorough-bred horse. The Revolution and the Civil War hindered Thorough-bred breeding but emphasized the need for fast, tough riding horses. John Morrissey opened his racetrack at Saratoga in 1864 and Pimlico opened its track in 1870, followed by the New Orleans' Fair Grounds in 1873 and Kentucky's Churchill Downs in 1875. The brothers Benjamin and Sanders Bruce were among the pioneers of the American Stud Book and produced a preliminary book in 1868.

Mr. J. S. Skinner established The American Turf Register in 1820. The American Stud Book was published in 1873 and ownership acquired by The Jockey Club (based in New York City) during 1896. Importation of Thoroughbred breeding and racing stock from Great Britain, France and Spain soon improved the quality of the racehorse in North America. Highly-bred stallions and mares were supplied from Great Britain to territories such as South Africa, Australia, New Zealand, Chile, Canada and Argentina.

The Godolphin Arabian (1724) was a brown horse standing 14.2 hands high with some white on the off hind pastern. After the demise of this stallion some horsemen argued he was not a purebred *Arabian* but rather a good example of the *Barb* in phenotype, but these criticisms were the result of examining portraits of the horse rather than seeing the real individual. In fact, he was a purebred Arabian of the Hajilfan strain from Yemen. He might have been presented to Louis XIV by the Emperor of Morocco, or stolen and shipped to Paris, but no matter which is truth, he was acquired by a Mr. Coke of Cambridgeshire, whose will bequeathed the stallion to a friend, Roger Williams. *The Godolphin Arabian* stood at Longford Stud, Derbyshire. Some years later he came into the ownership of the Earl of Godolphin, who bred Cade, sire of the influential stallion Matchem. Foaled in 1748, Matchem was a bay horse 15.1 hands high and was an excellent racehorse which retired sound from the racecourse when ten years of age. *The Godolphin Arabian* thus spent his remaining life at Brabraham Stud, Cambridgeshire until his death in 1753 at the age of 29 years. For much of his life this wonderful stallion was known by the nickname "Scham".

Among the progeny of *The Godolphin Arabian* which exerted substantial improvement on the development of the racehorse as a breed were Regulus, Janus, Mogul, Blank, Whitenose, Dismal and the famous mare Selima. Regulus sired successful stallion Fearnought as well as Spiletta, dam of famous sire Eclipse. Cade's male line descends via Melbourne, sire of West Australian and The Peer. West Australian was a triple crown winner and is the paternal ancestor of champion Man O'War (sire of champion War Admiral and War Relic) and of Display (sire of champion Discovery).

The male line of West Australian also survives via his son Solon (a roarer) which established Marcovil's branch of the line. Perhaps the very best sons of Marcovil were unbeaten Hurry On and Omar Khayyam. A daughter of West Australian produced tough stayer Musket, sire of champion Carbine, winner of the VRC Melbourne Cup with record weight. Marcovil's influence prevails via his famous daughter Miss Matty, dam of English Derby winner Papyrus, an ancestor which I consider to be of unique genetic importance in modern pedigrees of Group and Graded stakes winners. Papyrus carries recessive influence from two dynamic mares, Pocahontas and Angelica.

The Byerley Turk (a purebred Arabian) was imported to Britain in 1689 by Captain Byerley of County Durham and was used as an officer's mount in the Battle of the Boyne one year later. He did not receive much opportunity with quality mares yet sired Bonny Black, the grandest racemare of her era. The best son of *The Byerley Turk* was Jigg, sire of Partner. The male line was mainly carried on by Partner's grandson Herod, an important genetic influence on the breed for speed. Herod sired Highflyer, Florizel, and Woodpecker.

Florizel became the sire of Diomed, considered to be the foundation sire of the North American thoroughbred. Standing about 15.3 he won the initial English Derby and ten other races. After an indifferent stud career in England, Diomed was sent as a 22 year old to Petersburg, Virginia. His major sons were Truxton, Sir Archy and Duroc. Sir Archy is of course the grandsire of Boston, sire of the legendary Lexington as well as successful sire LeComte. Diomed's fame rested equally as a broodmare sire for his daughters produced many outstanding stakes winners in England and North America. Lexington was one of the first American stallions to transmit width of throatlatch (jaw space) and depth of chest.

Highflyer sired Sir Peter (actually registered as Sir Peter Teazle) whose male line descends via Paulowitz to Wild Dayrell, a very influential sire. Another branch comes via Sir Peter's son Walton, male ancestor of Macaroni, leading broodmare sire of England. However, the most dominant male branch of Herod came via Woodpecker and his son Buzzard.

Herod's line proved powerful and helped reshape the breed for classic speed. Buzzard sired two influential stallions, Castrel and Selim. These were elegant stallions with refinement but were hot-blooded animals.

Castrel is the male ancestor of unbeaten The Tetrarch, sire of Tetratema, Stefan the Great, Salmon Trout and champion 2YO filly Mumtaz Mahal. Selim is the male ancestor of The Flying Dutchman, Ksar (sire of Tourbillon), and Glencoe (sire of Pocahontas, and great grandsire of Hindoo, sire of Hanover) which transmitted short hind cannons to their progeny.

The Darley Arabian (a purebred Arabian) of 15 hands high by a *Mannicka* strain stallion from a *Managhi Hedraj* strain mare is considered to be the most prolific male line in the international Stud Book but is not necessarily the most influential in a genetic sense. From this line descend such modern sires as Gainsborough, Hyperion, Princequillo, Nearco, Nasrullah, Royal Charger, Fair Trial, Spearmint, Blandford, Teddy, Domino and Broomstick - all members of the Eclipse branch of *The Darley Arabian* male line.

A chestnut horse with superior speed and endurance, Eclipse established a male line which has dominated the breed. His son King Fergus established the line of Blacklock, sire of Voltaire, sire of Voltiguer, sire of Vedette, sire of Galopin. The best son of Galopin was unbeaten St.Simon, a bay horse which had an emphatic role in upgrading racing class. Vedette's son Speculum is male ancestor of Sundridge, sire of Sun Reigh and Sunstar.

St.Simon sired Rabelais (sire of Havresac II and male ancestor of unbeaten Ribot), Persimmon (male ancestor of Prince Rose, sire of Prince Chevalier and Princequillo), Chaucer (male ancestor of Vatout, Vatellor), and St.Serf (male ancestor of St.Frusquin and Desmond). St.Simon became a genetic force in the breed, substantially upgrading female lines with speed, but a legacy from his impact was to increase the chance for curbs in the hocks of his descendants. St. Simon sired the winners of 17 English classics. Foaled in 1700, *The Darley Arabian* was born near Palmyra,

secured when a four year old colt by Thomas Darley at Aleppo and sent to his father in Yorkshire, England, when six years old. He was about 15 hands high and came from the purebred strain known as *Mannicka*. A handsome stallion he quickly established a reputation and became the sire of the popular Flying Childers (great sire), Bartlett's Childers (great racehorse), Bulle Rock (first thoroughbred sire imported to North America), Dart, Brisk, Aleppo, and Almanzar.

Three of the most influential sires in thoroughbred history came from the Eclipse line of *The Darley Arabian*, namely Stockwell and his brother Rataplan, as well as their half-brother King Tom (broodmare sire of St. Simon).

Stockwell's progeny won three English Derbies, six St. Legers, four English 2,000 Guineas, three 1,000 Guineas and one English Oaks. He was described as the Emperor of Stallions and his son Doncaster established the male line of Bend Or, grandsire of Cyllene. This male branch is represented at time of writing by more representatives than any other sire line. Cyllene sired Polymelus, sire of Phalaris whose sons include Fairway, Pharos, Caerleon, Colorado, Manna, Sickle, Pharamond II and Colombo.

The Pharos male line is responsible for Nearco (sire of Dante, Sayajirao, Nearctic, Nasrullah, Royal Charger), Firdaussi and France's leading sire Pharis II (sire of Priam II). Fairway, full brother to Pharos, is represented in male line by Blue Peter and Fair Trial, the latter being sire of 2,000 Guineas winner Court Martial. Pharamond II became the sire of Menow, sire of champion Tom Fool; whereas Sickle begot Unbreakable, sire of Polynesian (sire of champion Native Dancer). Nearco, one of the all time great sires in the stud book, is grandsire of champion Northern Dancer whose sons include Storm Bird, Nijinsky II, Danzig, Dixieland Band. Another successful male line of Nearco comes via Nearctic's son Icecapade, sire of Clever Trick.

Bend Or's champion son Ormonde (like Bona Vista, produced from a mare by the famous stallion Macaroni) became the sire of Orme, one of the most brilliant racehorses of all time and produced from St. Simon's full sister Angelica. Orme won 14 of 18 starts including the Eclipse Stakes twice and was champion 2YO colt of England. Orme's male line continues via champion Flying Fox (grandsire of French champion Teddy) and via English Derby winner Orby (sire of The Boss and Grand Parade). Teddy in turn sired Sir Gallahad III, Bull Dog, Asterus, and Sun Teddy. Another strong male branch of Stockwell came through his son St. Albans, grandsire of the very influential stallion Sainfoin, an English Derby winner whose descendants helped shape the breed. I consider Sainfoin to be one of the most dynamically gifted ancestors in modern pedigrees. He is a *Genetic Giant.* His full sister Sierra produced champion sprinter Sundridge whose influence penetrates the breed in an alarmingly effective manner. English champion 2YO colt Nasrullah, leading sire in Britain and North America, is linebred to Sainfoin and his sister Sierra.

Sainfoin sired triple crown winner Rock Sand which in turn sired English champion Tracery (out of a daughter of Orme), Mahubah (dam of champion Man O'War, Mazda and My Play), Trap Rock (successful sire and brother to Tracery), and Friar Rock (sire of Pilate, sire of champion Eight Thirty). Tracery was sold for a huge sum of money to Argentina where he became leading sire. His best sons include Papyrus (English Derby), Copyright (sire of champion Congreve), Flamboyant, Abbot's Trace, Obliterate, Waygood, Tracer, Pantheon (a leading sire down under), tough racemare Teresina (dam of American speed sire Alibhai) and stakes producer Traverse (dam of Transmute and Traffic).

Via Birdcatcher's branch of Eclipse came the male line of Swynford, sire of Blandford which sired four English Derby winners including Blenheim (sire of Mahmoud, Mumtaz Begum and

Donatello II). Blandford remains a key player in the pedigrees of classic winners and is renowned for transmitting good knees to his descendants. If ever one wishes to correct flat knees or back-at-the-knee problems, one can do no better than seek descendants of Blandford. Oddly enough, upon researching Blandford's ancestry, I found his pedigree had an amazing amount of impact from a male ancestor. viz. the Leedes Arabian.

Bend Or also sired Radium, grandsire of New Zealand champion Phar Lap.

Finally we come to the Touchstone branch of *The Darley Arabian* male line. Touchstone became the sire of Orlando, male ancestor of Himyar and his famous son Domino; also of the outstanding racehorse Newminster, sire of Hermit (English Derby) and Lord Clifden (sire of Hampton, whose branch includes the line of Gainsborough, Solario, Hyperion, and Son in Law). Domino's line is perpetuated mainly via Alsab, male ancestor of champion Ack Ack and his son Broad Brush, but also via Bimelech's best son Better Self, Blue Larkspur and champion Stymie.

For historical purposes, I now supply what I regard as a fairly comprehensive list of the most influential stallions imported into Britain, that helped shape the racehorse. It has been stated by some authors of reference books on thoroughbred breeding that the gene pool of the British racehorse was quite small. My independent research shows that there were more than 200 stallions which covered mares prior to the importation of The Godolphin Arabian. Whilst the first General Stud Book lists a considerable number of broodmares, they all trace to about 100 taproot mares. However, only 30 of these sustained long-term impact, the remaining 70 taproots being of inferior quality.

Where possible I have given the recorded birth date of stallions, but in most instances I could only find the year of importation, so caution should be used when studying the following list.

EARLY STALLION IMPORTS OF NOTE

The Morocco Arabian (1596)
Markham Arabian (1606)
Lord Fairfax's Sultan (Arabian) (?) sire of Old Bald Peg
Cullen Arabian (?)
Massey's Black Barb (?) also known as Sir William Morgan's Black Barb
Fenwick Barb (?)
Sir Victor's Barb (?) sire of Bald Galloway
Lord Fairfax's Morocco Barb (Arabian) (1635)
Rutland's Barb (1650)
White Legged Lowther Barb (1670)
Oxford Dun Arabian (1715)
Belgrade Turk (Arabian) (1717)
Place's White Turk (Arabian) (1657) - also known as the Sedbury Turk
as well as Darcy's White Turk - owned by
Oliver Cromwell, whose studmaster was Mr. Place.
The Darcy Yellow Turk (Arabian) (1658) - also known as Darcy's Chestnut Arabian
Dodsworth's Turk (imported in utero) (1665)
Shaftesbury Turk (Arabian) (1680)
The (Old) Morocco Barb (Arabian) (circa 1680) - also called the Helmsley Turk, Bustler and the Taffolet Barb
The Lister Turk (Arabian) (1683)

The Byerley Turk (Arabian) (1684)
Oglethorpe Arabian (1689)
Leedes Arabian (1690)
Fenwick Barb (1690)
Pulleine's Chestnut Arabian (1690)
Selaby Turk (Arabian) (1695)
Curwen's Bay Barb (Arabian) (1697) - also known as Pelham's Barb
Holderness Turk (1704)
Chillaby Turk (1706)
Newcastle Turk (1697)
The Darley Arabian (1700)
Alcock's Arabian (1704) gray stallion also known as the Akaster Turk, Sutton Turk, Honeywood's Arabian, Sir J. William's Turk, Sir C. Turner's White Turk
The Belgrade Turk (Arabian) (1710)
Bethel's Arabian (1711)
Strickland's Turk (1712)
William's Arabian (1713)
Basset's Oxford Bloody Shouldered Arabian (1716) also called the Bassett Arabian
Wynn Arabian (1717)
The Godolphin Arabian (1724)

From tedious research of pedigrees, I discovered that nine of the above stallions collectively represent almost 40% of the gene pool of the modern European racehorse. In North American pedigrees, they represent almost 35% of the gene pool.

These nine stallions are Lord Fairfax's Sultan Arabian, Place's White Turk (an Arabian also known as Darcy's White Turk or the Sedbury Turk), The Byerley Turk (an Arabian), Leedes Arabian, The Morocco Barb (an Arabian also known as Bustler, Helmsley Turk, or Taffolet Barb), Curwen's Bay Barb (an Arabian also known as Pelham's Barb), The Darley Arabian, Bethel's Arabian, and The Godolphin Arabian. All nine were purebred Arabian stallions. The next most important stallion seems to have been the Alcock Arabian which went under various names.

DOMINANT ANCESTORS

During my 40 years study of the origins of the modern racehorse I analyzed the mixture of genes from various breeds of horses which comprised the initial gene pool of the thoroughbred. Here are my findings, using two separate tables:

TABLE A - Composition of Genotypes

Arabian and Turk (*Equus Arabicus*)	**70%**
Barb (whose ancestry is partly Arabian)	**18%**
Other - includes Galloways and Native breeds	**12%**

Total Original Gene Pool:	**100%**

Stallions and mares purchased in Turkey at Constantinople (Istanbul) were described as *Turks* even though the majority were purebred *Arabian,* for this was the accepted way to identify them from purebred horses purchased from Palestine and Arabia.

However, horses described as being *Barb* were not always pure, many being a mix of *Arabian* and *native* breeds. Genetically, *Barbs* carried the *Arabian* purity in lesser proportions so should never be classified equal to purebred animals. Using a reasonably large international sample of 1,200 winners my objective was to try to determine which ancestors played a decisive role in the development of racing class, so as to find the Top Ten foundation ancestors in modern pedigrees. This task took place during the 1980's. The most surprising feature to note by genealogists or breeding journalists is the inclusion of three mares in Table B:

TABLE B - Percentile of Influence, Top Ten Original Ancestors

1.	The Godolphin Arabian	12.7%
2.	(Old) Morocco Barb (Arabian)	8.3%
3.	Old Bald Peg (mare) and her sire Sultan (Arabian)	6.5%
4.	The Darley Arabian	6.0%
5.	Tregonwell's Natural Barb (mare) (Arabian)	5.5%
6.	Curwen's Bay Barb (Arabian)	5.5%
7.	The Byerley Turk (Arabian)	5.0%
8.	Place's White Turk/Darcy's White Turk (Arabian)	4.9%
9.	The Vintner Mare (Arabian)	3.8%
10.	Hautboy - often called Old Hautboy (Arabian)	2.5%

One would expect the sample to highlight the importance of the three famous stallions, i.e. *The Godolphin Arabian, Darley Arabian, Byerley Turk*. Who would have thought that the roan stallion Sultan owned by Lord Fairfax which sired famous mare Old Bald Peg would rate higher than two of the three famous stallions credited with greatest influence on the breed. I must admit, I was also surprised that the *Old Morocco Barb or Bustler* scored second highest with 8.3% share of the gene pool. He is never given full credit by historians.

Lady Wentworth was perceptive enough to mention the *Old Morocco Barb* in her book published earlier this century. Here is a rather informative passage worth quoting:

" The Morocco Barb of three hundred years ago was a pure Arabian, and so were the horses of Syria in the time of Rameses and till the eighteenth century. The immense Syrian studs B.C. and the great collections of King Solomon were typical of the scale on which good horses were bred in the East when there was nothing of the sort done elsewhere. In the Christian era we have Marco Polo's records of ten thousand Arabian horses exported yearly to India in the 13th Century, of the colossal distribution of Arabians in Turkomania by Tamerlane (1366-1405), followed by the systematic flooding of the Turkoman tribes with Arabian breeding stock by Nadir Shah (1687) whose enormous studs of thousands of the best mares were seized by the Turks and transferred to all kingdoms under Turkish rule." - Lady Wentworth

Another ancestor featuring in the Top Ten is *Curwen's Bay Barb,* a purebred Arabian stallion imported to England by a Roman Catholic Priest who was instructed by his cardinal to relinquish material ownership. The background of the stallion is sketchy, but from what I can research in 1698 Muley Ishmael presented to Louis IV of France a purebred Arabian colt and a filly. The Arabian filly was subsequently given as a gift to the English king, but the colt was given to an English priest as a token of appreciation for solving a political problem between the royal courts of

England and France. At the request of the priest's cardinal, *Curwen's Bay Barb* was handed over to the manager of the English king's Audley Stud. The horse later resided at Hampton Court. The filly was named *Curwen's Arabian Mare* while the colt was named *Curwen's Bay Barb*. When the stallion changed ownership he was known as *Pelham's Barb* and he represents an amazing 5.5% of the entire gene pool. He was a remarkable broodmare sire despite the fact his male line became extinct. Notice he is much more influential than *The Byerley Turk*.

Because the horse was worshipped and idolized by Bedouins, Christians, Hindoos and other religious groups, it was protocol to exclude mention of the horse in any religious writings or tablets. Perhaps this explains why the Old Testament rarely makes mention of the horse.

We now turn to brief notes of racing in England where the modern thoroughbred was actually developed. On the execution of King Charles 1st of England, Oliver Cromwell seized the Royal Studs in 1649. At the time these farms held 54 mares, 25 foals, 17 two year olds, 22 yearlings, 21 older horses and 20 stallions. Many of the breeding stock were sired by the purebred Arabian stallion Frisell (by the Markham Arabian) and the Barb stallion named Rupert, but 8 were also sired by the Black Morocco Barb. Nearly all were dispersed by private treaty in 1651.

Oliver Cromwell retained 6 stallions for riding purposes, stating that the rest were of substandard quality. Instead, he embarked on a policy of introducing some of the purebred Arabians available at the markets of Aleppo. In November of 1657, Cromwell's first imported horses arrived on the ship *Dartmouth* which anchored at Gravesend. Among these was the famous stallion called Place's White Turk, also known as Darcy's White Turk despite the fact the stallion was pure Arabian. A seven year old of supreme quality in the prime of his breeding career, Place's White Turk covered the best mares Mr. Place could possibly supply so as to upgrade Cromwell's cavalry. The stallion's progeny were tested in heats of up to eight miles.

Towards the end of the 17th Century, race meetings were held at Newcastle, Northampshire, Winchester and some other courses. From 1654 to 1658 Oliver Cromwell forbid thoroughbred racing for political reasons even though he was among the leading breeders of the racehorse.

A vital formative period of the development of the racehorse came with the Stuart restoration to the English throne in 1660, when King Charles II was crowned. After the restoration in 1665, King Charles II established the King's Prizes, known as the Royal Plates, run over the new round course at Newmarket with heats, under weights of about 12 stone.

These King's Plates played a significant role in testing racing stock and were a sounding board to spotlight the best strains. During the reign of King Charles II two catastrophes were to reduce the realm's human population, namely the Great Fire of London and the Great Plague. It was during the period 1660-1685 that some thirty Oriental mares and ten stallions were imported by the king.

These purebred Arabian mares and stallions were among the most important ancestors which helped establish the racehorse as a breed and were purchased at prohibitive prices. They formed the foundation material of the Royal Studs to help replace the destruction of Tutbury Stud which had been dispersed some years earlier. Most of the breeding stock at Tutbury Stud were sold to noblemen during the Rebellion. A week after his coronation, King Charles II appointed Lord D'Arcy as Master of the Royal Studs in 1661. The sixth son of the 7th Lord D'Arcy de Knayth of Hornby Castle in Yorkshire, James D'Arcy managed Sedbury Stud and was an animal breeder par excellence. One year later, the king was given as a present a number of Persian horses of pure Arabian ancestry, as well as many coach horses - gifts from European rulers. The king was an

excellent rider and sportsman and in Muir's English racing paper, *The Olde New-Markitt Calendar*, it quotes:

" Yesterday His Majesty (Charles II) rode himself three heates and a course, and wonn the Plate - all fower (four) were hard and neer ridden, and I do assure you the King wonn by good horsemanshipp."

James D'Arcy (born 1617) was responsible for the improvement of stock bred from the Royal Mares and it is partly due to his selection skills that tremendous upgrading of speed and stamina resulted in the next two decades. D'Arcy made a lasting impression with those original imported Arabian female lines which are shown in the first General Stud Book. In retrospect, King Charles II could not have selected a better manager for the Royal Stud's program. James D'Arcy married Isabel, daughter of Sir Marmaduke Wyvill who also owned a successful team of racehorses. It was common for stallions and mares to be named with the prefix Darcy, and so the Yellow Turk became known as the Darcy Chestnut Arabian. Under the reign of James II (1685-1688) gentlemen's races took place at Newmarket and in 1698 the same racetrack was visited by Russian Czar, Peter the Great, who purchased quality breeding stock to send back to his homeland. By 1887 the Royal Stud was producing quality thoroughbreds at Sandringham. Upon researching racing and breeding in England during the 19th Century I occasionally came across statements which exemplified the thinking of the times.

For example, John Osborne (born 1833) who rode his first race in 1846 made the statement:
" The method of training horses in the present day is quite different to what it was when I began. Horses are very much lighter now, they have neither the bone nor the substance that thoroughbreds had fifty years ago. Of course, in the old days they were sweated a good deal. Old John Scott was a great believer in sweating and bleeding horses."

Melton trainer Charles Lund gave a speech in 1899 which echoed these sentiments when he said:
" Present day trainers have gone in for speed too much, with the result that the great majority of the horses are too finely built."

ACCURACY OF EARLIEST RECORDS

The General Stud Book records are accepted as being fairly accurate, yet the first volume was not published until 1793 and information came from primary and secondary sources, some of which was either incomplete or inaccurate. Breeding records were difficult to ratify because often stallions and mares were given multiple names. For example, it may not have been clear to the publishers of the first English stud book that Place's White Turk was the same horse as the Sedbury Turk which was the same horse as Darcy's White Turk. Another good example is the Taffolet Barb (circa 1680) which went under the name of the Morocco Barb, Helmsley Turk and Bustler.

The great racehorse Flying Childers is given in the General Stud Book as being by The Darley Arabian from Betty Leedes by Old Careless (Spanker-a Barb mare) from Cream Cheeks by Leedes Arabian from a mare by Spanker from the Old Morocco Mare by Morocco Barb from Old Bald Peg by an Arabian stallion (Sultan) from an imported Barb mare. Readers should be made aware that Spanker was by the Darcy Yellow Turk out of the Old Morocco Mare. Therefore, the mare by Spanker in the pedigree of Flying Childers was inbred to the Old Morocco Mare (dam of Spanker himself). After much research, I have to agree with Lady Wentworth who believed this pedigree to be accurate. However, one of the foremost authorities on pedigrees of the English race-

horse was Mr. C.M.Prior. This historian attempted to raise doubts about the authenticity of the inbred mare in Flying Childers' pedigree, for he advocated strongly against inbreeding as a method of improving the breed. Mr. Prior claimed that a private stud book compiled by Cuthbert Routh suggested Flying Childers was by The Godolphin Arabian from Careless by Leedes Arabian from Sir Marmaduke' Wyvill's gray mare of unknown origin, but since Cuthbert Routh's private stud book contained many inaccuracies, its credibility is shot to pieces. During research of the pedigree and performance of Flying Childers and his full brother Bartlett's Childers (also known as Bartlett's Bleeding Childers) I suddenly realized here was the first true evidence of thoroughbreds known to have been "bleeders" on the racetrack.

One of the earliest stallions to impart speed and vigor to the development of the racehorse was a stallion named Greyhound, but I cannot find his date of birth. Other early stallions which helped shape the breed but whose influence is sketchy include Grasshopper and Warlock's Galloway, and the famous mare called Mr. Shield's Galloway, grand-dam of Tartar.

Lady Wentworth researched The Helmsley Turk, an Arabian import owned by the Duke of Buckingham and found he was the same horse as the Taffolet Barb, the Morocco Barb and Bustler, yet on page 513 of the General Stud Book it suggests the Helmsley Turk was the sire of Bustler. Lady Wentworth was quite sure they were one in the same stallion. Thus, early records of the stud book possessed flaws but one must remember that records were not always kept in good order and some information was indeed taken from private letters written by breeders from memory. Even private stud books contained inaccuracies, owing to the writer's lapse of memory. As a consequence, I am unconvinced that the three male lines in existence today reflect a true picture. There is every possibility that a fourth or fifth male line may have actually survived but mistakenly accredited to the three great stallions *The Darley Arabian, Byerley Turk and Godolphin Arabian.*

In 1980, the population of the thoroughbred exceeded one million. In March of 1978, Professors Garry A. Mahon and E.P. Cunningham complied a genetic study in an attempt to evaluate the importance of the three direct male line foundation sires. Their study was based on a small random sample of 60 mares drawn from Volume 35 of the General Stud Book. They hypothesized from their findings that almost 81% of genes in the British and Irish thoroughbred derived from only 31 original ancestors, suggesting that the genetic base of the racehorse might be even more narrower than they had presupposed. Their small random sample revealed the following percentages of the most dominant ancestors found in the pedigrees of the 60 mares under review:

MAHON and CUNNINGHAM STUDY (1978)

The Godolphin Arabian	14.6% share
The Darley Arabian	7.5% share
Curwen's Bay Barb	5.6% share
The Byerley Turk	4.8% share
Bethel's Arabian	3.3% share
Darcy's White Turk	3.3% share
Old Bald Peg (Arabian mare)	3.1% share
Sir Victor Barb	3.1% share
The Lister Turk	2.5% share
The Leedes Arabian	2.5% share

.........estimating 50.3% of genes for the above as a likely share of the gene pool.

Cunningham and Mahon subsequently indicated that two other important ancestors which influenced the breed were Mr. Curwen Shields' Galloway mare (grand-dam of Tartar) and Hautboy's dam. It might be of interest to compare their figures with those from my larger sample, covering a wider geographic source.

Garry Mahon's research for Professor Cunningham was published in Scientific American - May, 1991 issue. It was the first study to highlight the input of the forgotten stallion *Curwen's Bay Barb.* Patrick Cunningham was professor of animal genetics at Trinity College, Dublin, and has worked on genetic aspects of domestic livestock improvement for more than 25 years. He was director of the Irish National Agricultural Research Institute until 1988, but now works in Rome. He stated in the article: "Genetic studies of Thoroughbreds are particularly timely because of two problems facing the industry. First, Thoroughbreds display disturbingly low fertility: on a global average, only slightly more than 50 foals are produced by every 100 Thoroughbred mares. In the British and Irish population the annual reproduction rate is better - about 67 percent - but even this figure looks very poor compared with those in other single-offspring species, such as cattle, for which a rate of 85% is considered normal. The other problem is that contrary to what one would expect from the efforts to breed and train winners, the racing performance of Thoroughbreds is not uniformly improving. The three English classic races are the St.Leger, a 1.75 mile race open to horses of both sexes, and the Oaks and the Derby, 1.5 mile races open to fillies and colts respectively. Winning times for those races improved from the 1840's up to about 1910 but since then have been relatively static."

The Derby is also open to fillies as well as colts, and so too is the 2,000 Guineas. Professor Cunningham then goes on to state that perhaps the small gene pool of the thoroughbred might limit fecundity. In my opinion, the thoroughbred's original gene pool has enormous genetic variation and during the development of the racehorse from 1690 onwards, there were a multitude of modifications affecting the genetic codes of foundation stock. Valuable, as well as destructive gene mutations, took place on a gradual basis and today we find just as much genotypic variation in the breed as we would wish to cope with; so critics who grab the small gene pool concept are perhaps unaware of the variation which exists in the breed. Only professional people who work with thousands of thoroughbreds during a lifetime can truly understand what I mean. The fact is that the original gene pool is much *larger* than I ever suspected.

Rather than a base of about 80 individuals as is often quoted by historians, there were more than 500 individuals which played a role in the formation of the breed and this is why we have so much genotypic and phenotypic variation in today's racehorse. I have highlighted the Top Ten ancestors of supreme influence, yet one must not underestimate the other 490 individuals which shaped the breed. Their contribution enabled a degree of hybrid vigor to be maintained.

FOUNDATION MARES

The origins of many of the taproot or foundation mares whose families were given a Number to signify its importance by Australian pedigree researcher Bruce Lowe, is not entirely known. When the first English General Stud Book was published, insufficient pedigree information was available. Foundation mares came from different genetic sources and mainly comprised a mixture of different phenotypes. Many imported mares were of purebred *Arabian* ancestry, and about 25% were the result of blending the *Turk, Barb, European and Asian native horse* strains, including Galloways of uncertain origin. The *Arabian* influence was present in a quantity greater than was first recognized. As a result, initial development of the thoroughbred racehorse by British horse-

breeders during the 18th and 19th Centuries relied on the quality of pure-bred Arabian stock and very definite selection skills. Inbreeding took place simply because of transportation problems. Many celebrated stallions were inbred to superior females. With the passage of time and hindsight, leading British horse-breeders realized that maternal influence was all powerful.

In my previous book *Quest For a Classic Winner,* I provided charts of "Elite" mares from each of Bruce Lowe's Numbered Families. I explained how Bruce Lowe designated numbers to the best female lines according to each family's success rate in producing English Classic winners. The family producing the most classic winners became the Number 1 Family. The second best classic producing family became the Number 2 Family, etc. Whilst the numbers may have had some significance last century, today they are merely of historical value.

Successful female lines (families) rise and fall according to genetic probability and opportunity factors. *Nothing is constant except change itself.*

The horse-breeder of the 21st Century should have some inkling about the background of foundation mares classified with a Family Number designated by Bruce Lowe, because it is rather difficult to acquire this information owing to the rarity of old books. Here now are brief comments about those mares which were given Bruce Lowe Family Numbers.

FOUNDATION MARES AND FAMILY NUMBER

No 1 Family
Tregonwell's Natural Barb mare is a pure Arabian and was originally called Place's White Turk Mare and ran in races as Snorting Bess. She was born about 1657 and was a gray mare with a pronounced dish face very typical of the Arabian phenotype. This mare established the most important female line in the stud books of the world. She was imported to England from Aleppo and was a prolific winner of races. Her name was changed when Tregonwell Frampton managed the Royal Studs in the period following James D'Arcy's management. Famous broodmare La Troienne traces in direct female line to Tregonwell's Natural Barb mare and carries many other infusions throughout her pedigree.

No. 2 Family
Burton's Natural Barb mare was imported and her dam is believed to be pure *Koklana,* one of the best of the Arabian strains. It is possible that this mare may be related the foundation mares of Families No. 3, 17 and 22, as indicated by Lady Wentworth, but this is difficult to prove. Burton's Natural Barb mare produced two important daughters, an unnamed mare by Dicky Pierson (son of Dodsworth) and an unnamed mare by Dodsworth.

No. 3 Family
Byerley Turk mare named Treasure, dam of the Two True Blues mare which could trace to Burton's Natural Barb mare. The Byerley Turk mare was a daughter of the Selaby Turk from a mare by Darcy's White Turk.

No. 4 Family
Darcy's Layton Violet Grey Arabian mare. She was a very attractive gray mare by Darcy's White Turk from a Royal Mare by Layton's Grey Barb (also known as the Shaftesbury Turk). The female line traces to an imported Arabian mare. It is possible that Darcy's Layton Violet Grey Arabian mare is also the ancestress of Cutty Sark, a foundation mare in Australia.

No. 5 Family
Massey's Black Barb mare, by Massey's Black Barb from an Arabian mare imported from Egypt. She is believed to be almost pure Arabian and is sometimes referred to as The Massey Mare. She produced Old Ebony (by Bastro) which produced Ebony (by Flying Childers)

No. 6 Family
Old Bald Peg, by Lord Fairfax's imported Arabian stallion Sultan (of the Managhi strain) from the Old Morocco Mare sometimes called Old Peg which was pure Arabian and imported from Egypt. Old Bald Peg is the most influential mare in thoroughbred history and her sire was a roan chestnut of fine quality.

No. 7 Family
Darcy's Black Legged Royal Mare, also known as Darcy's Oldest Royal Mare named Whitefoot. She is the dam of Darcy's Old Sorrill (by Blacklegs) as well as an unnamed mare by Oxford's Dun Arabian which produced Miss Slamerkin (by Young True Blue), dam of Bustard.

No. 8 Family
Hutton's Grey Barb mare, tracing to the Bustler mare whose third dam is probably Darcy's Layton Violet Grey Arabian mare (see Family No. 4). By Hutton's Grey Barb (an Arabian stallion), she is out of an unnamed mare by Hutton's Royal Colt from a mare by Byerley Turk from a mare by Bustler.

No. 9 Family
The Vintner Mare, of pure Arabian ancestry, her dam supposedly imported directly from Aleppo. Considered to be a finely boned mare with true quality. She was by Curwen's Arabian. In my opinion she is the third most important foundation mare in thoroughbred history.

No. 10 Family
Childer's Mare, sometimes called the Oysterfoot Arabian mare and Miss Hip. She was out of Virgin by Ancaster Merlin from an unnamed Somerset mare from a Coppin mare from a Place's White Turk mare which could be related to Family No. 28. The Childer's Mare produced an unnamed mare by Gower Stallion, which in turn produced an unnamed mare by Snap. I am not convinced that the Childer's Mare is 100% Arabian.

No. 11 Family
The Sedbury Royal Mare also known by the name of Miss Darcy's Grey Wilkes or simply Old Wilkes. She was by Hautboy (sometimes referred to as Old Hautboy) a son of Darcy's White Turk, and produced from Miss Darcy's Pet Mare of pure Arabian ancestry. Champion racehorse St. Simon descends in direct female line from this mare.

No. 12 Family
Darcy's Royal Montagu Mare, by Darcy's Old Montagu from an unnamed mare by Hautboy from an unnamed mare by Brimmer, from an imported Royal Mare (probably from Aleppo). This mare is almost certainly of pure Arabian ancestry, but Lady Wentworth considered the possibility that the Royal Mare may also have been a daughter of Place's White Turk Mare. Sometimes Darcy's Royal Montagu Mare is referred to as the Old Montagu Mare, She is the grand-dam of champion racemare Spiletta, dam of champion Eclipse; and grand-dam of an unnamed mare by Whiteshirt which established a very successful line of stayers. The stallion Darcy's Old Montagu (often called Old Montagu) was imported to England by Ambassador E.W.Montagu from Constantinople about 1716, and that the stallion was a pure Arabian.

No. 13 Family
The Royal Mare, a chestnut Arabian import which might be related to the same female line of Families 7 and 11. She was among the expensive imports King Charles II purchased from Aleppo that was put in the care of James D'Arcy. She produced a filly by Darcy's Yellow Turk which in turn produced Grey Royal (by Darcy's White Turk), ancestress of Highflyer.

No 14 Family
The Oldfield Mare, by Sir Hales' Bay Turk. Thought to be about three-parts Arabian. She produced an unnamed mare by Hale's Turk which in turn produced an unnamed mare by St. Martin (Pelham's Arabian-Burton's Natural Barb mare) which produced Lady Mare (by Pert).

Lady Mare (by Pert) was sometimes referred to as Old Lady Mare. She produced an unnamed mare by Newton's Bay Arabian which became the sixth dam of famous mare Banter, dam of Touchstone. Champion mare Pretty Polly traces to this same female line.

No 15 Family
Grey Whynot, by Old Whynot (Fenwick Barb-a Royal Mare) from Darcy's Royal Mare imported by King Charles II from Aleppo, but unrelated to the Royal Mares above. I believe this mare was pure Arabian.

No. 16 Family
Sister to Stripling, by Hutton's Spot and tracing to a mare by Darcy's White Turk from a mare by the Holderness Turk. Research shows that there was a sister to Stripling whose grand-dam by The Darley Arabian traced back to the Oldfield Mare (see No. 14 Family). However, I am unconvinced this family is related to the Number 14 female line.

No.17 Family
Byerley Turk Mare, by the Byerley Turk from a mare imported to France via Constantinople. Believed to be of pure Arabian ancestry. She produced an unnamed mare by Bald Galloway which in turn produced The Wharton Mare (by Carlisle's Turk).

No. 18 Family
Old Woodcock Mare, by Bustler (otherwise known as the Morocco Barb or Helmsley Turk) from a mare by Davill's Old Woodcock. She is the same mare which Keylock's Charts erroneously show as being "Mare by Old Woodcock". The Old Woodcock Mare produced an unnamed mare by Old Spot which produced an unnamed mare by Curwen's Bay Barb.

No. 19 Family
Old Woodcock Mare, by Davill's Old Woodcock, but unrelated to the mare with the same name given in Family No. 18. Her dam is of unknown origin. She is ancestress of famous stallion Vedette.

No. 20 Family
Mare by Gascoigne's Foreign Horse, also known as Daffodil's Dam. She produced the high-class stayer Daffodil as well as Favourite (by Bald Galloway), dam of an unnamed mare by Little John.

No. 21 Family
Queen Anne's Moonah Barb mare, an imported Arabian mare from the Mananghi strain. This mare is sometimes referred to as The Moonah Mare. She produced an unnamed mare by Leedes Arabian, which in turn produced an unnamed mare by Old Whitefoot, which in turn produced an

unnamed mare by The Darley Arabian as well as an unnamed mare by Hobgoblin.

No. 22 Family
The Belgrade Turk mare, supposedly imported from Constantinople. I regard her as being pure Arabian. She produced an unnamed mare by Bay Bolton which produced an unnamed mare by Bartlett's Childers, which produced Flora (by Regulus).

No. 23 Family
The Hobby Mare, by The Lister Turk (purebred Arabian) from Lord Kingston's White Mare named Piping Peg, also by The Lister Turk. This means the Hobby Mare was intensely inbred. In my opinion Piping Peg is the source of white thoroughbreds with dark eyes, and her recessive genes have been transmitted in hidden form for many generations. Piping Peg was not a gray mare. She was an outstanding racemare and held in high regard because of her recessive white coat. She became the ancestress of the successful stallion Gimcrack, a stallion which sired a few white thoroughbreds during his stud career. The Hobby Mare is also ancestress of Solon and Barcaldine.

No. 24 Family
The Helmsley Turk Mare, by The Helsmley Turk (otherwise known as the Morocco Barb, Bustler or the Taffolet Barb). She produced an unnamed mare by Rockwood which in turn produced Old Lady (by Pulleine's Chestnut Arabian). She is the ancestress of influential sire Gohanna.

No. 25 Family
The Brimmer Mare, by Brimmer (The Yellow Turk-a Royal Mare). Dam of unknown origin but possibly related to the Number 4 female line.

No. 26 Family
The Merlin Mare, by Merlin. Possibly traces to the Oldfield Mare (see No. 4 Family) but it is impossible to substantiate.

No. 27 Family
The Spanker Mare, by Spanker. There is some evidence that this mare is a descendant of Old Bald Peg, foundation mare of the Number 6 Family. Lady Wentworth strongly suspected this idea.

No. 28 Family
The Coppin Mare, by Selaby Turk from a mare by Place's White Turk. Possibly traces back to either Tregonwell's Natural Barb mare or else Darcy's Layton Barb mare.

No. 29 Family
Arabian Mare, imported from Aleppo. She produced an unnamed mare by Basset's Arabian which in turn produced an unnamed mare by Bloody Shouldered Arabian. She is ancestress of Oaks winner Landscape.

No. 30 Family
Mare by the Duke of Chartre's Hawker of unknown female line origin. She produced an unnamed mare by Bristol's Hog which in turn produced Golden Locks (by Grasshopper) and his full sister.

Derby winning full brothers Archduke and Paris trace to this mare. This family has not been able to sustain itself due to poor fertility and disappointing performance. It is believed that the taproot mare carried very little Arabian influence.

No. 31 Family
Dick Burton's Barb mare, which was confiscated from Egypt. She produced an unnamed mare by the Chesterton Arabian which in turn produced the handsome stallion Woodcock (by Merlin) as well as an unnamed mare by The Darley Arabian.

No. 32 Family
Dodsworth's Dam, or otherwise known as the Royal Natural Barb Mare. She was imported from Tangier and is believed to be pure Arabian. She produced the high-class racehorse and successful sire Dodsworth as well as Vixen (by The Morocco Barb). Vixen in turn produced Old Child Mare (by Gresley's Arabian) which in turn produced an unnamed mare by The Darley Arabian. Dodsworth was an exceptionally fast stayer.

No. 33 Family
Sister to Honeycomb Punch, by the Morocco Barb (also called The Helmsley Turk, Bustler or Taffolet Barb) and sometimes referred to as the Bustler Mare. She produced an unnamed mare by Makeless which in turn produced Old Polly (by Flying Childers). While she is ancestress of Derby winner Sergeant, her line has become practically extinct.

No. 34 Family
Clubfoot, otherwise known as The Hautboy Mare, by Hautboy. Her dam is of unknown origin.

No. 35 Family
The Bustler Mare, by the Morocco Barb (also called The Helmsley Turk, Bustler or Taffolet Barb). She produced an unnamed mare by Byerley Turk which produced an unnamed mare by Bay Bolton. Her line has almost faded into obscurity.

No. 36 Family
Mare by Curwen's Bay Barb, tracing in direct female line to The Vintner Mare (see Family No. 9). She is believed to be almost pure Arabian in ancestry and is ancestress of River Prince.

No. 37 Family
Sister to Old Merlin, by the Old Morocco Barb from Old Merlin's Dam.

No. 38 Family
Twaite's Dun Mare, by the Akaster Turk (otherwise known as Alcock's Arabian). She produced an unnamed mare by Partner which is ancestress of Derby winner Sir Thomas and the remarkable racehorse Pot 8 O's, best son of champion Eclipse.

No. 39 Family
Bonny Black, by Black Hearty (son of The Byerley Turk) from an imported mare by a Persian stallion said to be from the Mananghi strain.

No. 40 Family
A Royal Mare, possibly related to Darcy's Old Montagu Mare (see Family No. 12) but I cannot substantiate this information.

No. 41 Family
Grasshopper Mare, by Grasshopper. She might be from the Highland Laddie mare which traces to the Oldfield Mare. Highland Laddie was full brother to Leeds, a very smart racehorse. This family is now obscure.

No. 42 Family
Mare by Spanker, dam of unknown origin. She produced an unnamed mare by Pulleine's Chestnut Arabian, but the family has not produced a decent runner for 40 years. Ancestress of Wychwood Abbot.

No. 43 Family
A Natural Barb Mare owned by Lord Arlington. She was bred by the Emperor of Morocco and is believed to have been a Barb, not of pure origin.

No. 44 Family
Mare by the Morocco Barb (also known as Bustler). A weak family of no modern importance.

No. 45 Family
Mare by Young Cade, ancestress of 1,000 Guineas winner Barcarolle.

Other Female Lines not given a Family Number by Bruce Lowe, many of which suffered by not being numbered, were:

Two mares by Brabham from different female lines
Spectator Mare
Mare by Sheild's Galloway (grand-dam of influential stallion Tartar)
Whitenose Mare
Miss Euston (sometimes called Miss Middleton by Middleton) and actually related to the No. 3 Family
Selima (1745) famous foundation mare in the USA, being by The Godolphin Arabian out of Shireborn by Hobgoblin out of a mare by Whitefoot.
Mare by Bolton (ancestress of Barbizon, Sword Dancer)
Gray Arabian Mare given to King William IV by the Imaum of Muscat
Sir John Sebright's Arabian Mare

The following mares were not entered in the General Stud Book, but were entered in private stud books some of which were researched by Miss C. B. Prior:

Mare by Roseden
Maggie Lauder (by Dinmont) - the family of Galloping Queen.
Mare by Perion
Meade's Celer Mare
Piersfied Mare
Tryall Mare, ancestress of Enquirer.
The Janus Mare, ancestress of Ben Brush, Bunting, Equestrian, Silver Fog, Silver Buck, Inca Queen, State Dinner and Kentucky Derby winner Gato del Sol
Cutty Sark Mare, an outstanding foundation mare in Australia
The Medley Mare, ancestress of English Derby winner Durbar II, Kentucky Derby winner Clyde Van Dusen, world champion Swaps, Courtly Dee and Green Desert
Mare by Vampire
Bess Brimmer Mare
Fearnought Mare, ancestress of Dobbins
Miss Bell
Fearnought Mare, ancestress of King Alfonso
Mare by Medley, ancestress of Spendthrift and Sweep On

Quicksilver Mare
Bowie Mare, ancestress of Irish Lad
Dinah Mare
Mare by Janus, ancestress of Novice (by Glencoe), dam of Norfolk and his sister The Nun, and third dam of Americus
Barb Mare, ancestress of King's Courier
Fearnought Mare, ancestress of winners in France
Gabriella (by Sir Archy)
Potomac Mare, ancestress of Emperor of Norfolk

It seems strange that one of the most influential mares of the 20th Century carries strains of Americus and Emperor of Norfolk in her pedigree, namely Lady Josephine (dam of Lady Juror and Mumtaz Mahal). It seems strange too that French champions Tourbillon and Djebel carry the genes of Durbar II, and that champion Swaps descends from a female line not considered worthy of representation in the first General Stud Book. I trust this brief summary of original foundation mares (which is incomplete) helps clarify the origins of the modern racehorse. Furthermore, other maternal influence coming via the dams of imported stallions is another factor to be considered.

In Frank Forrester's book published in 1857 there are indexed 28 Arabian stallions, 14 Barbs and a number of comments about Spanish horses. Francis Culver (1922) gives the story of General Harvey Lee and Lindsay's Arabian, stating that the Belsize Arabian was imported to Philadelphia, USA, in 1768. There were in fact many purebred Arabian horses imported into North and South Carolina, Virginia and Maryland from the late 1770's.

In France, a racecourse was established in 1776 on the Plaine des Sablons near Paris. Frenchmen had to wait until the end of the French Revolution of 1789 before thoroughbred racing was taken seriously. Emperor Napoleon created a series of races by a decree on August 31st, 1805, but it was an Englishman, Lord Henry Seymour and his friend the Duc d'Orleans (heir to the French throne) who encouraged the organization of French racing and breeding. These gentlemen founded the French Jockey Club in 1833 and shortly afterwards the Societe d'Encouragement was established. Also in 1776 the French nobleman, the Duc de Chartres, purchased quality thoroughbreds from England. The Marquis de Conflans followed suit. Racecourses at Longchamp and Chantilly were established.

During 1865, French-bred three year old Gladiateur (by Monarque) stunned British racegoers by winning the English Derby, 2,000 Guineas and St. Leger. Owned by Count Frederick de Legrange, Gladiateur became a symbol for French horse-breeding. The Grand Prix de Paris was an important new stakes race instituted in 1863, and open to foreign runners.

Many high-class British racehorses were exported to Argentina, Australia, New Zealand and South Africa, and the racehorse soon became an international athlete although quarantine regulations had to be introduced owing to viruses being transferred while in transit on ships. Later, Japanese and Indian buyers began to raid the sale-rings of Newmarket and Dublin.

The modern racehorse evolved from the mixture of a variety of different horse breeds and does not breed true-to-type like a pure-bred species. Racehorses come in all shapes and sizes, coat colors, rib and vertebrae numbers. Some still inherit annoying splint bones.

Head shapes differ (concave, convex), length of neck varies, shoulder angle has become more upright because of selection for early maturing speed, hocks have improved with less curbs than

DJEBEL, champion in France, son of French Derby winner Tourbillon.

MUMTAZ MAHAL (1921) owned by His Highness the Aga Khan with her colt foal (by Colorado) named Furrokh Siyar. Champion 2YO filly of England, Mumtaz Mahal (The Tetrarch-Lady Josephine by Sundridge) is the grand-dam of Mahmoud and Nasrullah.

say 100 years ago, but pasterns still vary in length. The shape of the croup has altered and tail-set is much lower. I am seeing more good racehorses with the important "triangle" in their hindquarters, denoting that they have sufficient length of hip bone.

Perhaps the most dramatic change has been size, for there are far more racehorses above 16 hands high, as a proportion of the breed than 100 years ago. Thoroughbred skeletons which were examined for purposes of osteology research, as given by Lady Wentworth in her book on page 375 show substantial variance. For example, Touchstone possessed 19 ribs:

	Ribs	Lumbar Vertebrae
Touchstone	19	5
Persimmon	18	5
St. Simon	18	5

compared with....

Eclipse	18	6
Hermit	18	6
Blacklock	18	6
Polymelus	17	6

Purebred Arabian strains consistently possessed 17 ribs, and mainly 6 Lumbar Vertebrae but there were instances whereby a few Arabian horses had 7 Lumbar Vertebrae.

Companies now service horse-breeders with equine measurements. We now have specialists in bio-mechanics which group thoroughbreds by specific conformational classifications (type) and claim success at being able to predict future racing class. Of course, accuracy in measuring racehorses is critical because data is entered into computers for reference; thus, any slight mistake in measuring a horse would endanger final prognostication of its athletic potential.

The racehorse breeder has a multitude of statistics to review all stallions and mares, more so than breeders of the past. Computerization has been able to quickly communicate race results within minutes, instantly supply pedigree extensions and scramble meaningful statistics for the racehorse breeder.

Statistics are provided for stallion fertility, Average Earnings Index, Comparative Index, Racing Index, number of stakes winners, percentage of winners to starters, stallion service fee fluctuations, average and mean statistics for yearling and weanling prices, speed ratings etc.

Advice can be sought from industry specialists such as the measuring people, pedigree consultants, sales company executives, consignors, veterinarians, trainers, jockeys and bloodstock agents. Their advice is of course subjective opinion because nobody as yet has access to the actual genetic composition of a racehorse, stallion or broodmare.

Curious observers of the thoroughbred can identify genetic characteristics or traits expressed in a dominant, semi-dominant or recessive manner. One has to use one's eyes in this business. In the final analysis, one has to admit that the vast majority of genes in the overall gene pool are *recessive*. Unfortunately, we know little about *Sex-Linked genes* and their impact on a foal but one doesn't need to be a brain surgeon to know they exist in the breed. *Sex-Linked genes* are found exclusively on the X sex-chromosome of a racehorse.

Rarely do we ever think about the presence of *lethal genes* in the breed, killer genes that are hidden and passed from generation to generation. When two recessive *lethal genes* meet up in a foal's genotype (one from the sire, one from the dam) they will be expressed and terminate the life of the foal. However, *lethal genes* are not expressed when inherited at single strength (from one parent only) and are masked until a certain mating enables another single strength killer gene to cause expression. Breeders of *Arabian* horses know that *lethal genes* exist in their breed, for many purebred foals have died from this problem.

Certain equine female lines transmit a multiplicity of ovarian cysts, seemingly inherited in a haphazard, recessive fashion. During the 18th and 19th Centuries blindness was a serious problem caused by harmful recessive genes but this problem seems to have been eradicated by the process of selection.

When thoroughbred foals are aborted (i.e. die in-utero) what is the actual cause? Is it a genetic mistake whereby Mother Nature terminates the pregnancy due to a hormone problem; or is it the expression of a *lethal gene?* Has *Arabian* input masked some unfavorable characteristics?

The Greek word "homos" means "the same" and in genetic terminology an animal is said to be *homozygous* if it carries two *alleles* of each gene and breeds true. This creates dominance because both *alleles* act the same. If the organism carries two different *alleles* it is said to be *heterzygous.*

Should a racehorse have superior ancestors common via both sire and dam, its chance of inheriting a degree of similar genes is much higher than that of an outbred racehorse. Rapid improvement was made by racehorse breeders in Britain using similarly related parents. The practice of inbreeding and linebreeding produced good and bad results depending on the quality of material used; but the "good results" seemed to breed more true-to-type individuals which became valuable breeding stock, in the long term. Similar instances of inbreeding and linebreeding to superior ancestors in other countries also produced high-class runners.

By 1916 it was known that genes are located in a linear sequence along chromosomes. The racehorse is a diploid organism with two complete sets of chromosomes in each cell and therefore two copies of each gene. The position of a gene in relation to other genes on the chromosome is known as the gene *locus*. A male horse has XY sex-chromosomes compared with a female's XX sex-chromosomes. A stallion may have valuable genes located on its X sex-chromosome which have no opposing gene *locus* on its Y sex-chromosome. With no gene to oppose it, the gene found on the X sex-chromosome can easily be expressed. This is how genetic influence can be passed on without change via the X sex-chromosome, and we call this *Sex-Linked inheritance*. It might explain part of the reason why some stallions are limited in their transmissive qualities, e.g. by only siring good fillies. I label this phenomenon "*Sex Bias*".

Mendelian inheritance was eagerly accepted by animal breeders and used as a tool to understand how genetic inheritance works. Whatever their function, genes are inherited by a foal according to the special genetic code established upon conception. A gene (the unit of inheritance) is a length of DNA which has a specific function, and it can be turned "on" or "off". The only exception are Mitochondria, found in the cytoplasm of cells. Mitochondria contain their own unique DNA - quite separate from the nuclear DNA found in the nucleus of cells. Therefore mitochondrial DNA is non-Mendelian.

This fact went unnoticed until the early 1980's and is a separate issue which is discussed in a later chapter.

For a species to adapt to new environments it requires flexible change, so Mother Nature provides change via the existence of *mutant* genes. When the sequences of *bases* in DNA is changed it is called a *mutation.*

Amino acids can be mistakenly altered to form new proteins and the resulting gene no longer is a perfect copy and instead, becomes *mutant.* If a mutant gene is lethal it can never be transmitted to the next generation; but when a mutant gene is useful, it is gradually incorporated over time into the organism's total gene pool. When various horse breeds were mixed to escalate equine evolution, mutational change was instigated. It was a vital tool for change.

If we could have access to genetic maps of great stallions such as St. Simon, Sainfoin, Hyperion, Blandford, Nearco, Native Dancer, Mr. Prospector, Buckpasser, Sir Tristram and Danehill we could perhaps identify special genetic differences with the gene maps of inferior stallions and learn from the analysis. Unfortunately this information is unavailable for comparison. However, we can observe the way genetic influence is passed down to subsequent generations, making note of any obvious dominant characteristics, and simultaneously noting the expression of any hidden recessive characteristics. A stallion's phenotype or appearance may play an insignificant role in the way he transmits genetic influence to his progeny. Secretariat failed to sire more than a few outstanding colts in his illustrious stud career and instead became renowned for his brilliant fillies like Lady's Secret, Terlingua, Bluebook, Fiesta Lady. By contrast, the plain stallion Danzig sired superior racing colts like Lure, Danehill, Pine Bluff, Chief's Crown which were much more appealing types compared with his own conformation.

The true worth of a stallion or mare is what it can transmit to its progeny. Only by test breeding can we learn what genetic gifts can pass on to the next generation by a specific stallion, or a specific mare.

Not always are superior athletes, champions of their generation, the best sires or best producers. Breeding the best to the best doesn't always succeed. Certainly we must use superior male and female racehorses as the desired breeding material to try to upgrade performance; however, until they can prove their worth via "transmissive qualities" they are almost as big a risk as lesser performed males and females.

Once proven at stud, stallions and mares should be eagerly sought and used. Even if they are old so long as they remain healthy they represent a far better risk than use of young unproven stallions or mares. History shows some of the best progeny of stallions were born in the twilight years of their father's stud career. This is not the case with mares aged above 20 years because by then they are worn out from having to bear many foals and nurturing them to weanling age.

Whereas a stallion might sire 70 or more foals per breeding season, a mare can only produce one foal in the same period of time. Therefore a mare's influence is more quickly dispersed into the gene pool via her sons which become stallions. Mares which produce more than one successful stallion are extremely valuable because their sons will continue to transmit maternal strength. One of the most famous mares in British thoroughbred history was Glencoe's daughter Pocahontas, dam of Stockwell, Rataplan and King Tom - two leading sires, three leading broodmare sires.

Furthermore, Pocahontas is ancestress of Wellingtonia, one of the most important ancestors found in the clever pedigree of champion Tracery - a stallion much admired by Federico Tesio, and an ancestor I love to duplicate in matings. Pocahontas (by Glencoe) possessed a *Genetic Code* of immense power - a genetic code capable of immediate upgrading for speed.

Selene (Chaucer-Serenissima by Minoru), a tough little mare which produced Hyperion, Pharamond II, Sickle, Hunter's Moon - all leading sires - plus the famous producer All Moonshine (dam of Mossborough), also must have inherited a special genetic code.

Plucky Liege (Spearmint-Concertina by St.Simon) dam of Sir Gallahad III, Bull Dog, Admiral Drake - all leading sires - as well as the fillies Marguerite de Valois (ancestress of Fappiano and Quiet American) and Noor Jahan (ancestress of Sanctus) is another wonderful mare which must have possessed a powerful genetic code.

Gene mutations involve the addition or deletion of one or a few *base* pairs or the substitution of one *base* pair for another. Mutations involving larger segments of chromosomes e.g. gene groupings, or whole chromosomes or sets of chromosomes are called chromosomal mutations.

The alteration or rearrangement of new *alleles* provides for superior or inferior adaptation. There is so much DNA in a cell that newly mutated alleles are not uncommon but rarely have a noticeable effect on the phenotype. Sometimes the sex-chromosomes are affected which decrease fertility or cause infertility. Normal males have XY sex-chromosomes and normal females XX.

When *abnormal* sperm fertilizes an ovum (egg) it sometimes will pass on its XY instead of either and the foal inherits a mistake, e.g. XXY sex-chromosomes. Alternately, a foal might inherit only one Y or X - a freak accident. Sadly, foals with YO die very young, but oddly enough, XO foals are known to have survived. "Super" females are those which inherit an abnormal hormone intake and exhibit masculine characteristics, causing them to be sub-fertile.

During *interphase* preceding meiosis, each chromosome (carrying thousands of genes) replicates itself and becomes two cromatids (two identical parts). The exchange of genetic material is called "*crossing over*" because genes from each homologue have crossed over to the other. Millions of cells divide. When mistakes occur at time of "*cross over*", genes will be rearranged or reshuffled.

Another interesting phenomenon is *Linkage*. Certain genes may be *linked* and defy separation for many generations. *Linked genes* are transmitted unchanged as a group to the next generation as a result of having strong chemical bonds.

The more one learns about inheritance, the more one understands why full siblings, i.e. full brothers and sisters, can look and perform unlike each other. Many breeders believe it is unwise to repeat a proven mating which has supplied a Group or Graded stakes winner. They think the best foal has already been produced.

This is not necessarily a good philosophy to keep because research of produce records of stakes producers indicates a successful mating has a good chance of repeating itself *providing both parents share common superior ancestors in the first five generations of the foal's pedigree.* When a stakes winner is the result of an outcross mating whereby no ancestor is duplicated within six removes of its pedigree, chances are that a repeat mating might fail. In this instance, the parents obviously possess wide genetic variation and it is *pure luck* if the best genes are inherited. Personally I believe one should definitely repeat successful matings when sire and dam have duplication of the same superior ancestors top and bottom in a stakes winner's pedigree.

When selecting a suitable stallion to help upgrade a mare's future foal, the tendency for a horse-breeder is to pay little attention to the grandparents of the foal. The parents seem always to be the main focus. With forty years experience, and study of thousands of matings, I am now convinced

The Aga Khan's MAHMOUD (Blenheim-Mah Mahal by Gainsborough-Mumtaz Mahal) wins the English Derby in record time.

HYPERION (Gainsborough-Selene by Chaucer) six times leading sire in England, won the English Derby Gr-1 and sired many classic winners including Owen Tudor. His pedigree shows inbreeding to St. Simon and full brothers The Palmer and Rosicrucian.

the four "GRANDPARENTS" possess more genetic impact on a foal's biomechanics and biochemistry, than its parents. In my opinion, the sire and dam of a foal act mostly as "carriers" or transporters of all genetic combinations possible via grandparents. This conclusion has been drawn from first-hand experience gained from breeding racehorses and other species, e.g. show-dogs and exhibition birds. However, it is obvious that the immune system of a foal is directly influenced by its dam.

Natural selection in the wild makes certain only the fittest will survive. The weak will be eliminated. Although man-made selection tolerates many genetic weaknesses in thoroughbred breeding, by and large, substantial progress was achieved during the past 100 years owing to careful selection of breeding material and judicious linebreeding to superior performers. Initially we owe a great debt to British horse-breeders. Next, we must recognize the improvement made in North America, Ireland, France and Italy with the upgraded British product. Now we must appreciate rapid progress which has been made in other countries such as Australia, New Zealand, South Africa, South America, and Japan.

GENETIC WEAKNESS

The existing gene pool of the modern racehorse contains an untold number of genetic faults and weaknesses. Horse-breeders are most reluctant to publicize genetic faults transmitted by a stallion or mare. It would be economic folly to contemplate making these facts public. In specific regions there exists dangerous recessive traits in breeding stock, e.g. genetic weaknesses causing bleeding from the lungs, left laryngeal paralysis, partial infertility, OCD lesions, soft palate and flapper abnormalities. These recessive traits are a threat to the survival of the breed's racing soundness - inherited in hidden form and unknowingly transmitted to future generations. When two recessives meet each other at the same *logus,* the specific trait will be exposed by expression.

Medication for horses which bleed might enable race clubs to provide sufficient numbers of starters for a meet, but if the industry tolerates this problem and remains short-sighted by mating breeding stock which are both known bleeders, what may I ask will this do to the future gene pool? Imagine the long term effect if a leading commercial thoroughbred stallion (covering huge numbers of mares during his stud career) transmits weak genes causing bleeding from the lungs. It might not be obvious he is transmitting a major problem, particularly if only 20% of his foals demonstrate bleeding. The danger lies with the remaining 80% of his foals which hide the recessive weakness and pass it on to future generations. We don't know exactly how much of the cause might be genetic, or whether it is nearly all stress related. Nevertheless, if we continue to breed foals from parents which are known bleeders we run a risk of perpetuating this weakness. Because recessive genes are hidden, they may go unnoticed for generations.

Sensible breeders know that no mating should ever be conceived when both parents have similar faults, but how do we know whether weak recessive genes are hidden in both parents selected for a future mating? I carefully study the characteristics of all grandparents because they give a clue as to what combinations of influence one might expect from a desired mating. The ideal match for any mare is to mate her with a stallion which can improve her faults - the objective being to UPGRADE her foal. Similarly, the ideal match for a stallion is to select a female which possesses strengths not found in that stallion's conformation. Among the most desirable requirements is correct shoulder angle. It's been fashionable for modern horse-breeders to match speed with speed, a practice often resulting in progeny with the wither placed too far forward. In recent years I have become conscious of a shoulder angle shift at recent North American, European and Australian yearling sales. For real fluent action, the shoulder angle of a racehorse needs to be set somewhere between a range of 45 to 48 degrees for energy to go unwasted and to ensure less stress is directed to the knees, ankles and pasterns. Upright shoulder angle puts continuous stress on an athlete's forelegs, especially pressure on sesamoids and tendons. Stakes runners possess correct shoulder angle, enabling fluent motion and supplying a chance for an over-stride. Manipulating the genes of a foal can be rewarding because the whole concept of breeding the racehorse is to own a better athlete than your competitor. One needs to know as much as possible about a stallion or mare's immediate ancestry in terms of its strengths, weaknesses and special characteristics. Some conformation faults do not lessen the chance of a runner winning stakes races because they do not affect soundness. Many Group 1 or Grade 1 winners inherit imperfect conformation yet in the main, stay sound in training. We see them in the winner's circle every month. However, certain faults of conformation can lead to unsoundness and there is nothing more frustrating than owning a colt or filly with Grade 1 ability which can never make it to the racetrack due to physical problems.

It is disappointing that less organized study has been made of the genetics of the thoroughbred than the genetics of other domesticated animals, the latter funded by companies in specific industries. The racehorse is the most documented domestic animal in the world, with pedigree records and racetrack results going back for more than 200 years. Racehorse breeders should have a single objective - to succeed on the racetrack - so one wonders why commercial stud farms have neglected scientific research. The fertility of the racehorse has been marginally improved by veterinary skills, yet it remains at an unacceptable level. Worldwide fertility of thoroughbred mares seems to fluctuate between 65-70%, whereas it is approximately 83-85% in breeds of cattle.

CONCLUSION

The modern racehorse possesses a genotype which has been modified through means of careful selection from original breeding stock entered in the first stud books of the world. The proportion of *Arabian* influence is overwhelming, yet its mixture with other equine strains produced a faster and bigger athlete. No longer can a purebred *Arabian* compete with today's thoroughbred, and it would be foolish to ever contemplate using *Arabian* breeding stock to upgrade for performance. Paleontologists estimated a problematical timespan for the evolution of the horse. Personally, I doubt their findings (guesswork) are anywhere near the truth of the matter. This statement is not made to invite controversy. I ask the reader to consider another viewpoint. Herewith is a comparison between my estimate and that of the paleontologists. You can be the judge:

EVOLUTION OF THE HORSE

	Paleontology:	(Average height)	Author's guess:
EOHIPPUS	55 million years	3 hands	13 million years
MESOHIPPUS	35 million years	6 hands	9 million years
MERYCHIPPUS	25 million years	10 hands	6 million years
PLIOHIPPUS	10 million years	12 hands	2 million years
EQUUS	Present	16 hands	Present

Already controversy exists between geneticists and paleontologists about the evolution of man. In the book *"A Monkey Puzzle - A Family Tree"* written by John Gribbin and Jeremy Cherfas, (published by Triad in 1983) the authors postulate how studies of the DNA of the Mitochondria of cells established that man's origins are probably more recent than paleontologists would like to admit.

Humans, gorillas and chimpanzees shared a common ancestor almost 5 million years ago according to specific DNA examination, rather than the separation of 20 million years suggested by most, but not all, paleontologists. Genetic studies also show that there is only a 1% differentiation in their genetic codes between humans, gorillas and chimpanzees - i.e., their DNA!

Of course the 1% difference carries vital change in phenotype and intelligence and represents many thousands of genes, but like it or not, it must come as a shock to humans to learn that we have 99% of genes in common with apes.

After studying the evolution of the horse, I believe genetic research will one day confirm that my belief about the development of Equus (the modern horse) is more acceptable than that of paleontologists whose guesswork could not incorporate genetic study. The development of Equus is much more recent than we are told by text books. Whilst a considerable amount of subjectivity is used on my part to arrive at conclusions, it will be fascinating for my descendants to see if in time I guessed correctly. Research into the DNA of Mitochondria can confirm or deny the guesswork because the non-Mendelian DNA of the Mitochondria rarely changes, whereas the nuclear DNA constantly changes and can't be used as a guide. Not until genetic studies of the mitochondrial DNA of the racehorse are conducted shall we know the facts. In my opinion, the guesswork by paleontologists is wide of the mark by many millions of years.

DANEHILL (Danzig-Razyana by His Majesty), Group winner in England, and leading sire in Australia. Bred by Juddmonte Farms, he is inbred 3 x 3to Natalma, dam of the great Northern Dancer.

The late Stavros Niarchos, owner of the Haras de Fresnay-Le-Buffard in France, receives one of the many trophies his homebreds have won. He is pictured with his charming daughter Maria.

GREAT ANCESTORS WHICH SHAPED THE BREED

> " Racing is a matter of the heart. The poorest punter is blood brother, for a few moments as the race climaxes, to the millionaire owner. A third factor makes the racehorse special. A work of art - a Rembrandt, or a Turner, a genuine Chippendale sideboard, a Queen Anne silver cup can all afford comparable pleasure to the eye and to the investment portfolio. They can all be beautiful hedges against inflation. Their capital value may soar. But they cannot earn. They cannot, like the successful racehorse, breed."
>
> \- Ivor Herbert

Animal breeders in general need to acquire knowledge about the great males and females of a species or breed in order to be able to capitalize on their superior genetic influence. No other animal has had more documentation than the thoroughbred horse and the stud books of the world contain its heritage.

When DNA analysis becomes a reality, detailed gene maps will portray or define each type of gene representing the biochemistry of a racehorse. Wealthy racehorse breeders who can afford DNA information will have a distinct advantage over those who cannot afford such information. If gene groupings for heart size, lung capacity, electrolytic production, muscle fiber etc. can be identified in stallion or mare, it follows that by simple comparison Genetic Engineers will know which foals actually inherit the most desirable genes from their parents.

Inheritance would be traced and measured from an individual's gene map. However, traits of persona such as "courage" and "will-to-win" may still remain a mystery. For the interests of all horse-breeders I hope it does. There are hundreds of thousands of racehorses shown in the Stud Books of the world, yet how many of these really count? How many ancestors possess highly valuable genotypes which might be reinforced in future generations?

Stakes winners represent less than 5% of all runners in a given racing year and of this percentile not all are used for breeding purposes - some are gelded, some die, some are infertile. To improve performance, an animal breeder requires quality breeding stock with which to experiment and upgrade. The objective should not be to improve the whole horse population or herd quality; instead a racehorse breeder should be aiming to design matings for individualistic quality, to be highly competitive against the best runners in a crop.

To be competitive on the racetrack, an owner's racing team requires a few high-class athletes for Group or Graded stakes. Therefore, the breeder must first establish a breeding program which incorporates use of superior stallions and quality mares from stakes producing damlines and reinforce the better elements of their pedigrees. Some of the best performed racehorses at Group or Graded level possess less fashionable pedigrees and their stud fee or season price is often within the grasp of small breeders with a limited investment. One needs to be pragmatic when setting objectives for any breeding and racing program.

To meet this goal, the breeder must ask himself or herself:

a) "Where am I now?"

b) "Where could I be?"

c) "How do I get there?"

Setting objectives is easy, achieving them is difficult. Try to be honest and ask yourself are your goals realistic, affordable and achievable? First, it is necessary for a racehorse breeder to acquire knowledge about *superior* stallions and mares which transmit *genotypes* essential for high-class speed and stamina, no matter whether your aim is to breed allowance handicap runners or Group or Graded stakes winners. The higher one sets one's breeding program *standard* the more favorable should be the results, dependent of course on one's financial budget which can be an evil constraint. Naturally, it takes time to acquire knowledge about ancestors in pedigrees and newcomers should seek professional help before spending their first dollar. You need to know whether a stallion or mare transmits genetic faults and what lurks within their ancestry. Slowly but surely one has to acquire knowledge about conformation and temperament.

I encourage everyone to treat thoroughbred breeding as a business rather than a hobby. Don't leave your brains on the gate-post as you enter this industry, but make sure you enjoy what you are doing. It is a fascinating business with exciting challenges. Along the way you will come in contact with a kaleidoscope of personalities .... craftsmen and comics, gentle folk and dreamers, tough guys and tricksters, speculators and conservatives, royalty and rags. Always remember, everyone can be humbled by the horse-breeding business.

I began to study the evolution of the racehorse when I was ten and can't tell you why this passion developed, but I can admit to being a curious kid. The subject of "inheritance" captured my imagination and attention. When I was only seven, my elder brother Bill and I bred budgerigars *(small parakeets)* for exhibition purposes so we could make pocket money. We studied the principles of *Mendelism* and read English books about how to breed budgerigars for exhibition, using the method of inbreeding or linebreeding to the very best cocks and hens in our aviary. This was a quick way to upgrade breeding stock. Gradually we learned the basics of genetics by studying each generation of young birds we bred, noting any improvement of type, color, feather quality, headskull size, etc. Our exercise with parakeets was rewarding because it taught me the workings of hidden qualities carried by breeding stock - the annoying *recessive* factors.

Observing the transmission of *recessive* traits by generation was fascinating. When the best cock was matched with the best hen, it sometimes supplied disappointing results for they failed to breed true-to-type. Budgerigars come in various colors, wing patterns and feather texture, so we had plenty of variables to manipulate. A few years later the racehorse became my passion. I went back to basics and studied pedigrees of the stock which formed the nucleus of the thoroughbred gene pool. I'd read where famous Italian horse trainer Federico Tesio, breeder of unbeaten champions Nearco and Ribot had studied the origins of the breed, so if it was helpful to him, I determined it must be helpful to me.

I traced families for multiple generations, wrote out six generation pedigrees of famous racehorses bred in each decade from about 1800 onwards, searching for any "pattern" that might indicate how one could upgrade for racing class. I had to learn about the characteristics of stallions and mares which possessed extraordinary prepotency.

I was trapped! My mind became mesmerized by pedigrees and performance. Thus when I turned twelve, I took a trolly ride on a path of genealogical research and never applied the brakes. None of my family were interested in thoroughbreds so they failed to understand reasons for my mammoth hours of study of racehorse breeding.

During my learning curve I became more and more interested in genetics. Mendel was just as much a boyhood hero to me as was Superman, Batman, Tarzan and of course my favorite movie star, Brigitte Bardot. Mendel's theories and discoveries consumed my spare time. Why? Well my brother and I had successfully applied Mendel's principles in breeding exhibition birds and we made lucrative pocket money from the exercise. I read articles and books published on breeding the racehorse, absorbing facts and opinions - many of those books purchased at second-hand bookstores. I studied photographs of stakes winners to learn more about their conformation. Rather than accepting the opinion of racing journalists, I questioned everything.

To keep an open mind when researching a subject is not only fundamental but also of paramount importance. I was taught at school to always try to maintain an open mind. As a result, I was equipped to form my own opinion about ancestors in pedigrees after about 30 years study of the subject. Knowledge was gained slowly but surely.

Perhaps the most fascinating horse-breeding article which triggered a positive response in me was one published in "The British Racehorse" describing the work of a clever Frenchman, Colonel Vuillier. This pedigree analyst was responsible for planning racehorse matings for the Aga Khan, and since the Aga rose to become leading breeder and owner in England with homebred products such as Bahram, Blenheim, Mahmoud, Tulyar and Nasrullah, I wondered whether Colonel Vuillier had hit on something important, or whether success had been achieved simply because of the strength of the Aga's broodmares and the opportunity to match them with Britain's best sire material.

From meticulous pedigree research, Colonel Vuillier determined that only a few ancestors possess and transmit *superior* genetic influence - powerful influence which might be transmitted to future generations. He reasoned that by far the majority of thoroughbreds were of undesirable quality, with conformation weakness, and believed one should concentrate solely on the top 1% of the breed to be ultra competitive. Vuillier set this objective for the Aga Khan's breeding program and it proved extremely successful. He concentrated on the best ancestors for his client's breeding program and succeeded.

After an exhaustive study of all three year old classic winners in Britain, Vuillier concluded that success could be achieved by copying the ancestral *ingredients* of these classic winners. In his studies, he identified numerous important sires but only one important mare, namely Stockwell's dam Pocahontas. Those superior males plus the one mare Pocahontas represented common denominators found in pedigrees of all English classic winners.

Vuillier developed a unique system whereby these powerful ancestors were given numerical *dosages (mathematical units)* which could be used as a *Standard for Dosage*. He designed matings to ensure the numerical *dosage* for every powerful ancestor approached the ideal mathematical dosage shown from the pedigrees of classic winners. For example, if one of the Aga Khan's mares was deficient in the amount of influence of St. Simon's sire Galopin, he would seek stallions which had an abundance of Galopin's input so that the foal's *dosage* would come closer to the *Standard for Dosage* he desired for Galopin. Similarly, figures for other powerful ancestors were examined until the right stallion was found to suit the mare. Because of Vuillier's success for his wealthy employer, I reasoned his system had merit (even if for the wrong reasons stated in the Vuillier article). Champion racehorses resulted from use of "dosages" and people were inquiring about it. The first two *Series* of Vuillier's "dosage" system were revealed by articles published in The British Racehorse. After Colonel Vuillier passed away, the Aga Khan persisted with the system, convincing the gracious but elderly Madame Vuillier to continue her husband's work.

The Vuillier dosage system attracted the attention of Italian lawyer, Dr. Franco Varola, who was a contributing European correspondent to The British Racehorse. An author who now lives in Brazil, Dr. Varola appreciated Colonel and Madame Vuillier's work and introduced a more updated version, but whereas the Vuilliers used mathematics in determining *dosage*, Varola tossed aside any reference to Galton's Law of diminishing influence by generation and identified key ancestors which should appear in classic pedigrees. Thus, there were no numbers to calculate for Dr. Varola may not have believed one could breed high-class horses by any figure system.

With subjective judgment, Varola listed key sires and labeled them "Chefs de Race" yet he decided to ignore female influence, thus not one mare formed part of his exercise, which in retrospect, was a great pity.

An intelligent academic, Varola promoted his concept of "Chefs de Race". It soon attracted a cult following in Europe. Delighted with initial response and amused by pressure from his newfound disciples, Varola defined the list of stallions into specific categories according to the most appropriate distance range, viz. Brilliant, Intermediate, Classic, and Stout. This made it easy for people to quickly identify an ancestor's input in terms of speed and stamina. I do not wish to encourage horse-breeders to play the dosage game and shall refrain from going into detail about "Chefs de Race", yet it is important that I mention the concept because it was borrowed by a Californian named Dr. Steven Roman who adapted Varola's concept into what became the dosage profile which appears on computer printouts in North America.

It was promoted by popular *Daily Racing Form* journalist Leon Rassmussen. Dr. Steven Roman re-introduced mathematics to establish his revised dosage system. Personally, I find it of no use whatsoever. Logic prevents me from ever supporting a mathematical system to breed a good racehorse. Anyone with a general knowledge of genetics would find it hard to swallow. You can have two full brothers with the same dosage, yet their performance on the racetrack can be astronomically different.

However, there is always some value from research, and via the deeds of Dr. Franco Varola and Dr. Steven Roman we can perhaps examine important males which have shaped the breed. Of course subjective opinion was used to identify the new "Chefs de Race" which were re-classified. Unfortunately, the exercise concentrated on only half the available information. Varola and Roman neglected to include any great females (mares).

Meaningful research does not omit half the sample; hence, we must learn the hard way and try to identify valuable male and female ancestors which transmitted positive influence. A starting point is to review the history of *Leading Sires of Winners* as well as tables showing the best sires of two year olds, and the best broodmare sires. Probably the next point is to examine the pedigrees of Group 1 and Grade 1 stakes winners (especially classic winners) so as to familiarize oneself with any ancestors common to the make-up of these stakes winners. One can find this information in bloodstock libraries, or via the purchase of magazines and books which could start one's personal library. Perhaps the next step would be to collect old Stallion Registers which detail extended pedigrees, performance, and female line information about all commercial stallions. You will be amazed at the knowledge one can gain from reading these registers. They are chock full of information. Next, begin to collect new and old sales catalogs which contain data for individual broodmares, including their dams, grand-dams and great grand-dams. Major yearling sales catalogs should never be thrown away. Use them for reference. Go to the horse sales and learn to observe the different phenotypes which belongs to specific strains.

If one doesn't have sufficient time or the inclination to study pedigrees, I suggest you hire a specialist who enjoys this work. I cannot stress the importance of learning the value of *Maternal Influence* because after more than forty years intensive study on the methods of breeding the racehorse, the most fundamental aspect so crucial for success is to acquire broodmares from the best stakes producing families. Fillies and mares from the best families are always sought by the top breeders. Around the globe you will see the most successful farm operators active at acquiring quality mares and fillies. They know maternal strength is the pathway to success.

BLUSHING GROOM, champion French 2YO and champion miler.

Early in the 20th Century, the Aga Khan was shrewd enough to ask English trainer George Lambton to purchase exquisitely bred yearling fillies to establish his breeding nursery. The choice of a man so experienced with bloodlines like George Lambton to help buy yearlings was to set the Aga Khan on a successful path. Lambton selected many outstanding fillies for the Aga Khan including brilliant English Champion 2YO filly Mumtaz Mahal (The Tetrarch-Lady Josephine by Sundridge) which established an incredibly successful branch of the Number 9 Family.

From Mumtaz Mahal, the Aga Khan was able to breed stakes winners with amazing speed. Mumtaz Mahal produced brilliant colts like Badruddin and Mirza II. Her daughters included Mah Mahal (dam of Mahmoud and Mah Iran) as well as Mumtaz Begum (dam of champion Nasrullah, one of the most powerful sires of the century, and Sun Princess which produced Royal Charger and Tessa Gillian). Australia's champion 3YO colt of 1996 is Octagonal, tracing in direct female line to Tessa Gillian.

When I was hired in the latter part of 1985 by William T. Young to help establish Overbrook Farm, one of my roles was to seek mares from sound stakes-producing families to establish a strong genetic base for the farm's future. Among mares and weanlings purchased were European champion filly Mrs. Penny (dam of Northern Park which topped the Keeneland sale as a yearling at $2.8 million), Buzz My Bell (dam of Kentucky Derby winner Grindstone), Seaside Attraction (winner of the Kentucky Oaks, and dam of champion 2YO filly Golden Attraction and Red Carnival), Always Mint (dam of Mountain Cat, richest earning two year old in North America), My Yellow Bird (dam of Canadian 2YO champion Blue Finn and of Fancy Ribbons), Sea Sands (dam of Future Storm), Stellarette (dam of Hollywood Starlet Stakes winner Cuddles) and others. My other role was to help attract suitable mares for a young stallion called Storm Cat, and for the first three crops we tried to give this stallion mares with compatible genotypes. This homework paid off, but it took time to show results.

The objective was to acquire fillies and mares from strong stakes producing families. Without strong foundation mares for a program, one may never acquire a team of competitive runners at Graded stakes level. Maternal strength is vital for success in the horse-breeding business. One has to compete with powerful racing stables like the Maktoum's stable in Europe, Allen Paulson and Ogden Phipps in North America, and breeders like Daniel Wildenstein and Maria Niarchos in France. So, to be competitive, one needs to breed from mares related to high-class runners.

When a racehorse weighing about 500 kilograms pumps approximately 250 liters of blood through its system each minute when running at top speed, this might represent almost ten times its total blood volume. Beyond six furlongs, it is the level of lactic acid in the bloodstream which affects the speed of an equine athlete, and the racehorse must clear the rapid build-up of lactic acid to sustain speed beyond a mile. Stallions and mares with *higher clearance rates of lactic acid* should be identified and used as future breeding material. The horse's weight plus the jockey puts tremendous pressure on the delicate bones found in the ankles and pasterns of a runner's forelegs, so one needs to select for soundness as well as speed.

Many fast, but unsound colts and fillies have proven to be successful at stud, so one must weigh up the pros and cons of an individual for breeding purposes. There are no rules that one can make or follow in this game.

To help young horse-breeders identify the most powerful influences in the stud books of the world, I wish to give my subjective opinion about the *Genetic Giants* of the breed. The following list has been carefully made after considering performance and transmissive qualities.

GENETIC GIANTS OF THE BREED

PHASE 1

Stallions:		*Mares:*	
Sultan	(?)	Place's White Turk	(?)
Spanker	(?)	Old Bald Peg	(?)
Partner	(1718)	Julia (by Blank)	(1758)
Squirt	(1732)	Giantess	(1759)
Regulus	(1734)	Prunella	(1788)
Blank	(1740)	Penelope	(1798)
Matchem	(1748)	Mare by Alexander	(1790) - dam of Selim, Castrel, Rubens
Herod	(1768)		
Eclipse	(1764)	Evelina	(1791)
Highflyer	(1774)	The Cub Mare	(1762)
Sir Peter	(1784)	Tramp's dam	(1803)
Whiskey	(1789)		
Waxy	(1790)		

PHASE 2

Stallions:		*Mares:*	
Blacklock	(1814)	Ada	(1824)
Pantaloon	(1824)	Banter	(1826)
Voltaire	(1828)	Beeswing	(1833)
Glencoe	(1831)	Pocahontas	(1837)
Birdcatcher	(1833)	Martha Lynn	(1837)
Bay Middleton	(1832)	Diversion	(1838)
Melbourne	(1834)	Reel	(1838)

PHASE 3

Stallions:		*Mares:*	
Newminster	(1848)	Queen Mary	(1843)
Stockwell	(1849)	Brown Bess	(1844)
Rataplan	(1850)	Ellen Horne	(1844)
Lexington	(1850)	Maid of Masham	(1845)
Macaroni	(1860)	Woodbine	(1850)
		Miss Agnes	(1850)
		Sandal	(1861)
		Rouge Rose	(1865)
		Maggie B.B.	(1867)

PHASE 4

Stallions:		*Mares:*	
Galopin	(1872)	Quiver	(1872)
Hampton	(1872)	Mannie Grey	(1874)
Hermit	(1874)	Angelica	(1879)
Isonomy	(1875)	Black Duchess	(1886)
Man O'War	(1880)	Sierra	(1889)
St. Simon	(1881)	Mrs. Butterwick	(1890)
Sainfoin	(1887)	Paradoxical	(1891)
Orme	(1889)	Canterbury Pilgrim	(1893)
Domino	(1891)	Chelandry	(1894)
Cyllene	(1895)	Fairy Gold	(1896)
Flying Fox	(1896)	Pink Domino	(1897)
William The Third	(1898)	Festa	(1898)
Sundridge	(1898)		

PHASE 5

Stallions:		*Mares:*	
Chaucer	(1900)	Sunshot	(1901)
Rabelais	(1900)	Gondolette	(1902)
Rock Sand	(1900)	Frizette	(1906)
Broomstick	(1901)	Mahubah	(1910)
Ajax	(1901)	Plucky Liege	(1912)
Sweep	(1907)	Lady Josephine	(1912)
Sunstar	(1908)	Cinna	(1917)
Tracery	(1909)	Pearl Maiden	(1918)
La Farina	(1911)	Selene	(1919)
The Tetrarch	(1911)	Black Ray	(1919)
Teddy	(1913)	Marguerite	(1920)
Ksar	(1918)	Uganda	(1921)
Copyright	(1918)	Mumtaz Mahal	(1921)
Blandford	(1919)	La Troienne	(1926)
Pharos	(1920)	Nogara	(1928)
Papyrus	(1920)		
Sir Gallahad III	(1920)		
Asterus	(1923)		
Blue Larkspur	(1926)		
Bull Dog	(1927)		
Tourbillon	(1928)		

Notice that Phase 5 has a considerably high representation compared with Phase 4. During the 1900 to 1929 period in the development of the thoroughbred, the population of the breed began to escalate. The number of foals in France, Italy, Germany, and North America saw rapid increases but of course World War II took its toll on numbers in the 1939-1945 six year span in Europe.

PHASE 6

Stallions:		*Mares:*	
Hyperion	(1930)	Lavendula II	(1930)
Umidwar	(1931)	Astronomie	(1932)
Discovery	(1931)	Bloodroot	(1932)
Fair Trial	(1932)	Myrtlewood	(1932)
Mahmoud	(1933)	Feola	(1933)
Precipitation	(1933)	Djezima	(1933)
Mieuxce	(1933)	Boola Brook	(1937)
Nearco	(1935)	Boudoir II	(1938)
Panorama	(1936)	Perfume II	(1938)
Djebel	(1937)	Majideh	(1939)
Nasrullah	(1940)	The Squaw II	(1939)
Princequillo	(1940)	Arietta	(1941)
Wild Risk	(1940)	Flaming Top	(1941)
Court Martial	(1942)	Cutaway	(1943)
Royal Charger	(1942)	Papila	(1943)
Polynesian	(1942)	Rough Shod II	(1944)
Petition	(1944)		
Tudor Minstrel	(1944)		

PHASE 7

Stallions:		*Mares:*	
Better Self	(1945)	Grey Flight	(1945)
My Babu	(1945)	Past Eight	(1945)
Star Kingdom	(1946)	Almahmoud	(1947)
Summertime	(1946)	Striking	(1947)
Sicambre	(1948)	Bold Irish	(1948)
Tom Fool	(1949)	Tessa Gillian	(1950)
Native Dancer	(1950)	Little Hut	(1952)
Turn-To	(1951)	High Voltage	(1952)
Ribot	(1952)	Mixed Marriage	(1952)
Swaps	(1952)	Somethingroyal	(1952)
Traffic Judge	(1952)	Aspidistra	(1954)
Klairon	(1952)	Queen's Statute	(1954)
Bold Ruler	(1954)	Cequillo	(1959)
Crepello	(1954)	Aimee	(1957)
		Missy Baba	(1958)
		Key Bridge	(1959)

PHASE 8

Stallions:		*Mares:*	
Never Bend	(1960)	Noblesse	(1960)
Northern Dancer	(1961)	Ciboulette	(1961)
Buckpasser	(1963)	Tamerett	(1962)
Graustark	(1963)	Natashka	(1963)
Damascus	(1964)	Glad Rags II	(1963)
Sharpen Up	(1964)	Intriguing	(1964)
Biscay	(1965)	Mira Femme	(1964)
Caro	(1967)	Neriad	(1964)
Nijinsky II	(1967)	Miss Swapsco	(1965)
His Majesty	(1968)	Native Partner	(1966)
Hoist the Flag	(1968)	Riverside	(1966)
Mill Reef	(1988)	Courtly Dee	(1968)
Halo	(1969)		
Riverman	(1969)		
Roberto	(1969)		
Lyphard	(1969)		

PHASE 9 (incomplete)

Stallions:		*Mares:*	
Mr. Prospector	(1970)	Ballade	(1972)
Sir Tristram	(1971)	Delsy	(1972)
Blushing Groom	(1971)	No Class	(1974)
Northern Taste	(1971)	Grecian Banner	(1974)
Irish River	(1976)	Eight Carat	(1975)
Danzig	(1977)	Kamar	(1976)
Nureyev	(1977)	Fall Aspen	(1976)
Storm Bird	(1978)	Height of Fashion	(1979)
Sadler's Wells	(1981)	Never Knock	(1979)
Rainbow Quest	(1981)	Maximova	(1980)
Danehill	(1986)	Coup de Folie	(1982)
Sunday Silence	(1986)		

How were these stallions and mares selected? Put simply, they represent breeding stock which were capable of supplying high quality genotypes, and in the author's opinion were capable of upgrading racing class to Group 1 or Grade 1 level. A few of the stallions mentioned were not leading sires but became extremely influential broodmare sires. Phase 9 is incomplete and the reader may wish to add to this list by the end of the century.

Many triple crown winners or champion racehorses have not lived up to expectation as sire material. Citation, one of the all time greats in North America, sired only a few progeny anywhere near his own ability, yet he had ample opportunities to do so. Gay Crusader, an English triple crown winner, is another example which disappointed as a sire, yet he later proved to be a successful broodmare sire. Another problem with male lines is that whereas some stallions made the leading sires' list with their progeny, their sons failed to maintain that success in the next generation. Why do these phenomena occur? Why is it difficult for champions to sire progeny of similar ability? Genetic probability factors govern every mating, thus we can understand that a reshuffling of genes can give different results: but perhaps the answer lies with Mother Nature who sometimes causes performance to regress, or revert to the "average" of the population.

Another plausible reason is that some champion racehorses inherit their ability from the lucky combination of thousands of single strength genes which happen to be expressed in a positive way for racing class, but when the stallion transmits his genes via sperm these single strength genes are reshuffled and do not link up with similar single strength genes of the same quality. High-class performers which fail to produce progeny with above average performance are deemed to lack prepotency. What is prepotency? A prepotent stallion or mare is one which can be matched with a variety of different strains and yet can consistently produce quality athletes. Prepotent stallions possess genotypes (genetic constitutions) which are transmitted via dominant genes; thus, their progeny resemble each other in appearance (phenotype). Simply put, they dominate other strains and their physical features are visible in their progeny.

Stallions and mares shown on the list of *Genetic Giants* have proven their worth as being thoroughbreds capable of upgrading performance not only via their progeny but also via many of their descendants. At the beginning of the development of the breed, Partner and Regulus were stallions which could be relied upon to upgrade speed over a mile and a half, and the stallion Blank became an extraordinary success as a broodmare sire. Let's take a look at the pedigree of Partner (1718) whose sire Jigg was inbred to the Old Morocco Mare, a daughter of famous mare Old Bald Peg:

			Arabian sire	
		Byerley Turk		
			Arabian mare	
	Jigg			Darcy's Yellow Turk
			Spanker	*Old Morocco Mare*
		mare		Morocco Barb
PARTNER (1718)			*Old Morocco Mare*	Old Bald Peg
			Arabian sire	
		Curwen's Bay Barb		
			Arabian mare	
	mare			
			Old Spot	
		mare		White Legged Lowther Barb
			mare	The Vintner Mare

Partner's sire Jigg was a very influential sire of winners and while he was slightly temperamental he did upgrade for speed. Partner's broodmare sire Curwen's Bay Barb was one of the most influential sires of his time but possessed some sex-bias, siring far better fillies than colts. Here is the first example of a stallion which might have possessed sex-linked characteristics. Partner's sons were successful sires, and he sired a son named Tartar, produced from a mare by the good performer Fox. Tartar also became a leading broodmare sire so perhaps he too inherited certain sex-linked characteristics. In the pedigree of highly successful sire Gohanna (1790), I noticed he is linebred to Curwen's Bay Barb as well as to Squirt, Partner, Tartar and Mogul. It might be relevant to show this pedigree:

				Squirt
			Marske	The Ruby Mare
		Eclipse		Regulus
			Spiletta	Mother Western
	Mercury			*Partner*
			Tartar	*Meliora*
		mare		*Mogul*
			mare	mare (*)
GOHANNA				*Partner*
(1790)			*Tartar*	*Meliora*
		Herod		Blaze
			Cypron	Selima
	mare			Cade
			Matchem	mare by *Partner*
		Maiden		*Squirt*
			Pratt's mare	mare by *Mogul*

Gohanna carries Curwen's Bay Barb via Partner (three times) and via the mare shown as (*) which was by Sweepstakes from a mare by Bay Bolton from a mare by Curwen's Bay Barb. Not shown in the pedigree is further linebreeding to Bay Bolton, another successful sire. Gohanna was a very good racehorse, and is one of the first genetic examples of combining all three important sires of the mid 18th century, namely Eclipse, Herod and Matchem. Gohanna has a well balanced pedigree pattern for he carries duplication of superior ancestors via sons and daughters which may have increased his prepotency levels.

One of the early genetic affinities (or nicks) was to match Herod with mares carrying the strain of Matchem. This cross duplicated Partner via a son and a daughter, and seemed to upgrade for speed. Gohanna played a significant role in the development of the breed, and in later generations his strain had an affinity with that of successful sire Cade (Godolphin Arabian-Roxana) whose third dam Cream Cheeks happened to be a half-sister to Jigg, thus reinforcing the strain of Spanker's dam. I shall not dwell too much on the pedigrees of the great sires and mares of the 18th century, but I do wish to make a few comments about the three great sires Eclipse, Herod, and Matchem, for they maintained their sire lines up until the present time.

Eclipse was by Marske out of the outstanding racemare Spiletta, a mare inbred to Snake and Hautboy. Eclipse's sire Marske was by Squirt whose unnamed dam is by Snake out Grey Wilkes by Hautboy. Squirt's dam happens to be full sister to the grand-dam of Regulus, sire of Spiletta, hence the champion racehorse Eclipse is the product of inbreeding to a son and daughter of the same mare. It is possible that the maternal strength from this inbreeding gave Eclipse some prepotency. The pedigree shows a triple duplication of Snake as well as Hautboy, but credit must also be given to 5 sources of the Lister Turk and 9 strains of Darcy's White Turk in Eclipse's pedigree, which probably has more genetic effect than the closer single strain of The Darley Arabian in his make-up.

Herod (Tartar-Cypron by Blaze) is basically an outbred stallion with no common ancestors to my knowledge. His second dam was Selima, a mare by Bethel's Arabian, said to be a purebred Arabian horse of immense quality. Herod's sire Tartar is one of the best sons of Partner. In the stud book, Herod was officially named King Herod, but it has become common practice to call him Herod.

Matchem (Cade-mare by Partner) is inbred to the dam of Jigg and Cream Cheeks, and since his dam is by the influential Partner, he must surely have inherited much of Partner's speed. His sire Cade was by The Godolphin Arabian from Roxana by Bald Galloway from a mare by Akaster Turk from Cream Cheeks by Leedes Arabian from a mare by Spanker. Matchem's direct female line was probably superior to that of either Eclipse or Herod, in a genetic sense.

Why have I mentioned these old pedigrees? All three stallions Eclipse, Herod and Matchem helped shape the breed and were formidable upgraders or *Genetic Giants*. For prepotency to exist, a stallion or mare should have the best genes reinforced on both sides of their pedigree (through sire and dam). When superior genes are reinforced in a pedigree, they allow for the expression of common traits (characteristics) and breed more true-to-type. Matchem's foals were quite similar in conformation probably because of his inbreeding. Eclipse came into a similar category. Herod, an outcross, did not stamp his foals. In the case of Gohanna, he became famous as the broodmare sire of Tramp, which in turn became the broodmare sire of Glencoe, one of the greatest sires ever to stand at stud in North America. Glencoe sired only a small number of foals in England before his export to the USA, but among them was the elite mare Pocahontas, dam of Stockwell, Rataplan, King Tom. Glencoe became leading sire in North America and also leading broodmare sire. In his pedigree we find a triple duplication of Herod and Highflyer (both superior sires) and the combination of two 3/4 genetic siblings, namely Bacchante's dam, and Gohanna , which were both by Mercury from mares by Herod. Glencoe happens to trace in female line to Prunella, best daughter of Highflyer, a particularly strong maternal line which produced many classic winners in England. Inbreeding is a technique which can be used skillfully to reinforce certain genetic qualities that need to be recaptured for future generations. By duplicating an ancestor quite close up in a pedigree, the objective is to collect the duplicated ancestor's best genes at *double strength*. When two parents each carry similar genes at *single strength* they each contribute a sameness of the genetic code and so a foal from these parents can inherit characteristics at *double strength*. To show by example what is possible, here is the pedigree of English Derby and English Oaks winner Eleanor, a filly which was superior to all the best three year old colts of her generation:

ELEANOR	Whiskey	Saltram	Eclipse
			Virago by Snap
		Calash	Herod
			Teresa by *Matchem*
	Young Giantess	Diomed	Florizel
			mare by Spectator
		Giantess	*Matchem*
			Molly Longlegs

This classic winning filly is inbred 4 x 3 to Matchem, a stallion whose pedigree shows inbreeding to the dam of both Jigg and Cream Cheeks. Eleanor's sire Whiskey is bred on a similar pedigree pattern to Gohanna, possessing the major strains of Eclipse, Herod and Matchem.

Eleanor's full sister Julia produced Phantom, winner of the English Derby. The family's value was further advanced when Priam won the 1830 English Derby, he being by Emilius out of Cressida , a full sister to Eleanor. These examples help show how inbreeding to superior ancestors can have a lasting, positive effect.

INBREEDING

When we discuss the terms *Inbreeding* or *Linebreeding* we refer to the closeness of ancestors being duplicated. If the same ancestor in the top half (sire) and bottom half (dam) of a pedigree appears within two, three or four generations, we call this closeness "inbreeding". If it occurs from the fourth generation onwards, i.e. further removed in the pedigree, we call this "line-breeding". Each generation of a horse's pedigree is called a "remove". Hence, we would refer to the inbreeding to duplicated ancestor Matchem as being in the fourth remove (in the sire) and third remove (of the dam), or simply say, inbred 4 x 3. Linebreeding is the same as inbreeding except that it takes place further back in the pedigree.

The term *Inbreeding* was derived from the old saying of "breeding in to specific ancestors" so as to gain strong influence from the duplication of ancestors. The reason why breeders chose to *inbreed*, was to "fix" characteristics of the duplicated ancestor in future foals. Breeders did not wish to dilute valuable influence from highly prized stallions or mares, so they *inbred* to them to ensure their dominance would persist. However, some breeders were timid about the method of *inbreeding*, believing it to be an un-Christian behavior, and so they chose to duplicate superior ancestors further back in the pedigree, from the fourth generation onwards. This method of more distant inbreeding was called *Linebreeding*.

When we refer to racehorses as being full brothers, or full brother and sister, or full sisters, we are describing thoroughbreds which have the same sire and the same dam. Traditional ways of describing horses has always referred to progeny of the dam rather than the sire. Hence, if we were to describe Cahill Road and Unbridled (both by Fappiano-Gana Facil) we would call them full brothers (full siblings) but when their dam produces a foal by a different sire, we call it a half-brother (if a colt), or a half-sister (if a filly). The emphasis is always on the dam (the mother) for relationship. When we refer to three-quarter brothers, it means they share three parts of their pedigree in common, even if their sire happens to be the same horse.

For example, Nasrullah is three-quarter brother to Royal Charger. In this instance, we can see they are by the same sire from closely related mares. Nasrullah is by Nearco from Mumtaz Begum by Blenheim from Mumtaz Mahal. His three-quarter brother Royal Charger is by Nearco from Sun Princess by Solario from Mumtaz Begum.

Three-quarter siblings can have the same dam but have different sires. For example, speed sire Badruddin which became leading sire in Argentina, is by Blandford out of Mumtaz Mahal. By comparison, Nasrullah's dam Mumtaz Begum is by Blenheim (son of Blandford) out of Mumtaz Mahal. Thus, Badruddin is three-quarter brother to Mumtaz Begum.

When yearlings by a stallion are sold at auction, we do not refer to them as half-brothers or sisters even though they are, in a strict genetic sense. Two colts by Mr. Prospector are not called half-brothers. Two colts by Danzig are not called half-brothers. To be related, they must always have a relationship *through their dam.* There are instances whereby two yearlings may be by the same sire but produced from full sisters. These were called "brothers-in-blood" (if colts), "sisters-in-blood" (if fillies), or "brother and sister-in blood" (if both sexes).

However, the word "blood" was traditionally used prior to the science of genetics. It is a carry-over from the word bloodstock because in past centuries it was thought characteristics were transmitted via the blood, rather than genes. I prefer to substitute the prefix "genetic" to describe these closely related yearlings. Thus, a brother-in-blood should be called a "genetic brother" in modern terminology, and a sister-in-blood called a "genetic sister".

ST. SIMON

The most dramatic impact for improvement of racing class materialized in England with the advent of a "super" horse named St. Simon (Galopin-St.Angela by King Tom), a bay colt foaled in 1881 when his dam was 16 years old. Bred by Prince Batthyany, St. Simon went undefeated in 9 races during two racing seasons and became the leading sire in England between 1890 to 1896 inclusive, 1990 to 1991, and was leading sire of broodmares between 1903 and 1907, and again in 1916. Nominated for the Derby, St. Simon became ineligible upon the death of his owner-breeder, but among his most important victories were the Ascot Gold Cup and Goodwood Cup. He was an exceptionally versatile performer winning three races over 5 furlongs as a juvenile, and up to 2 miles and 5 furlongs, the distance of the Goodwood Cup.

A temperamental horse with an impressive physique, St. Simon possessed an equal measure of the three vital muscle fibers one seeks in a great racehorse - fast, medium and slow twitch fibers. He inherited extremely strong hindquarters, yet his hocks showed signs of "curbs". The most notable progeny of unbeaten St.Simon were Persimmon, St. Frusquin, Diamond Jubilee, Rabelais, Chaucer, Concertina (dam of Plucky Liege), Memoir, and he is often found as the duplicated ancestor in great sires and mares. Nearco and Hyperion are linebred to St.Simon and probably inherited much of his racing class. They too, became *Genetic Giants*. I am unable to substantiate a belief I have about St. Simon, but it is possible he was the original source of what we today call "sickle-shaped" hocks.

Many of St. Simon's descendants inherited sickle-shaped hocks, particularly Nearco's high-class son Nearctic, sire of champion Northern Dancer. The pedigree of Nearctic shows him to be by unbeaten Nearco from Lady Angela by champion Hyperion. Nearco and Hyperion are both linebred to St. Simon. Whether the shape of the hindlegs enable St. Simon's descendants to inherit a longer stride is something I have considered as a distinct possibility. Conformation experts consider sickle-shaped hocks to be a minor fault. In hindsight, I am uncertain whether it is really a fault, or merely a modification of structure.

The modern racehorse evolved from a reasonably limited gene pool yet within it were a multiplicity of genetic variables. Hence, it is via careful selection of breeding stock that some thoroughbred breeders have been able to be more successful with their runners on the racetrack than other breeders. My research of the most famous thoroughbred breeders over the centuries has enabled me to form the opinion that their success came by acquiring females from the best stakes producing families, and matching them with proven sires.

Competitors in Group 1/Grade 1 races happen to be the best athletes of their generation, and whereas last century great horses would win by many lengths, today this happens rarely because the talents of the best stakes winners might be only one second faster than say a Group 3 competitor. Improvement of the thoroughbred's speed will be dealt with in a later chapter. What we need to appreciate is the fact that although there are many unknown factors in breeding stock, caused by hidden recessive genes, we must focus upon the superior slice of the gene pool, the top 5% cream of each crop. Only by testing one's foals on the racetrack can we determine whether progress can be achieved and the selection process is reliant upon racetrack performance. Overheard at a recent Keeneland sale was a comment I've never forgotten. The guy in question was a commercial breeder and he whispered to his wife: " Darlin, I don't care if the colt never wins a race, so long as we get top money, 'cause I bred this one to make the catalog page appealing enough to big buyers". I squirmed and walked away.

The list of *Genetic Giants* might help readers because I am sharing my 40 years experience by making such a list. The more modern stallions and mares represent powerful ancestry and should be used for purposes of duplication in future pedigrees. The reader should impose his or her own beliefs and add any other high-class stallion or mare to the list, providing they are of top quality.

Special stallions and mares on the *Genetic Giants* list require separate mention. Superiority of racing power doesn't necessarily lie in the more perfect symmetry or condition, because inferior-looking or small horses often vanquish the more attractive, bigger opponents. A difference might be *vitality* and/or *nervous energy,* but essential to a superior athlete is a large strong heart, clean air-flow, all three vital muscle fibers, strong bone, intelligence, courage and soundness. These features enable the thoroughbred to have more speed. While conformation faults cause wasted energy, and decrease fluent action, racehorses with faults can still compete with the more perfect specimens of the breed at distance less than one mile.

In Phase 1 of the tables, Matchem, Herod and Eclipse were responsible for substantial improvement of conformation in their progeny, but not necessarily improvement of temperament. They sired "hot-blooded" runners with above average nervous energy. Of the mares listed in Phase 1, first place goes to the pure Arabian-bred mare Old Bald Peg, a dynamic female whose genetic influence has penetrated the breed worldwide. Prunella is also a mare with a big heart.

In Phase 2 of the tables, Melbourne is significant because he was one of the most useful stallions to increase size, and stood 16 hands high whereas his contemporaries were between 15.1 hands high and 15.3 hands high. Melbourne was influential in supplying good bone density (especially under the knees) and improved size. He possessed a slightly higher wither than Voltaire or Blacklock. Of the mares in this phase, mention should be made of three incredibly dominant mares, which not only upgraded racing class with their progeny, but enabled through their descendants for sire lines to survive. These three mares were Banter, Beeswing and Pocahontas.

By the time Phase 3 was completed, the racehorse had almost ended a rapid escalation of improvement in terms of speed and stamina, temperament and soundness. Stockwell was a sensational sire of classic winners and high-class stayers. He was so popular that he became known as the "Emperor of Stallions". Stockwell was a chestnut horse foaled in 1849 by The Baron (a highly nervous horse) from Pocahontas (a roarer) by Glencoe and was the fifth foal of his dam. He had a blaze on his face and stood almost 16.3 hands high with a slightly roman-nosed head, thereby depicting genetic input from his *Barb* ancestry rather than the dish-faced nose normally found in horses of *Arabian* ancestry. The most critical aspect of his conformation was the shortness of his fore and hind cannon bones. This valuable feature for soundness was transmitted on a regular basis to his progeny. Stockwell won 16 races during four seasons of racing but did not win as a juvenile. At three, he won the English 2,000 Guineas, Great Yorkshire Stakes and English St.

Leger, but went unplaced in the Derby in which he was the victim of severe interference. He became leading sire seven times and was a very influential broodmare sire. Famous American stallion Lexington was responsible for upgrading for speed in North America, and had an affinity with mares carrying Glencoe's strain. Macaroni was one of the greatest broodmare sires in thoroughbred history. Of mares in Phase 3, Queen Mary and Sandal are standouts. Readers should note Woodbine was a daughter of Newminster's full sister Honeysuckle. When Woodbine's descendants met up with Newminster's descendants it caused linebreeding to a brother and sister combination which resulted in the production of numerous classic winners. Maggie B. B. was one of North America's most significant foundation mares.

Phase 4 saw the appearance of the phenomenal St. Simon, but there were other stallions of supreme genetic importance like the small but courageous stayer Hampton, the nervous but talented Isonomy, Derby winner Sainfoin which would have a dramatic effect on classic speed via his descendants, and another sensational broodmare sire, Derby winner Hermit, a known bleeder. Also in this list was North America's champion racehorse and sire Man O'War, and the speed oriented stallions Domino, Orme and Sundridge. Orme was out of Angelica, a full sister to St. Simon. Of the mares listed in Phase 4, mention should be made of English Oaks winner Canterbury Pilgrim which enabled two sire lines to be perpetuated (via her sons Swynford and Chaucer) and Black Duchess, whose impact is forever more remembered via her descendants Blandford, Teddy and Bayardo.

Phase 5 begins with representatives born in the 20th century. In North America, the exploits of the full brothers Sir Gallahad III and Bull Dog are exceptional, for they improved the quality and speed of the Kentucky thoroughbred immeasurably after the huge impact of Broomstick. In France, St. Simon's son Rabelais became a sensational sire and broodmare sire because he was fortunate enough to cover mares free of his sire's genes. Rabelais is of course the sire of Havresac II (broodmare sire of Nearco) and of Rialto, sire of Wild Risk. Ksar, Teddy and Tourbillon were other formidable stallions in France, whose descendants penetrate international pedigrees.

Phase 5 is also significant because Blandford sired four English Derby winners, Pharos sired Nearco and Pharis II, and unbeaten The Tetrarch transmitted speed and classic speed. Blandford, Pharos and The Tetrarch are found in the pedigree of one of the all-time great stallions of the 20th century, Nasrullah. Of the mares in Phase 5, Selene became a standout by producing four influential sires in Hyperion, Sickle, Pharamond II, Hunter's Moon as well as two influential daughters in All Moonshine (ancestress of Sir Tristram) and New Moon. Plucky Liege is another fantastic mare whose sons include English Derby winner Bois Roussel, French 2,000 Guineas winner Sir Gallahad III, speed sire Bull Dog, and the Grand Prix de Paris winner Admiral Drake (sire of the dam of champion Turn-To). Lady Josephine became a source of brilliance and produced The Recorder as well as champion Mumtaz Mahal and Lady Juror. Mumtaz Mahal established a dynasty for the Aga Khan's studs and her descendants include Nasrullah, Royal Charger, Mahmoud, Tessa Gillian. Lady Juror produced Sansonnet (dam of champion Tudor Minstrel) and leading sire Fair Trial.

Phase 6 was affected by World War II. In this phase are stallions which sired the very highest standard of performance at a mile to a mile and a half. Later, emphasis was switched from stamina to speed and the evolution of the racehorse took a different turn. The popularity of two year old racing persuaded thoroughbred breeders to use sprinting strains, thus stallions like Fair Trial, Nasrullah, Royal Charger, Petition, Panorama, Tudor Minstrel and Court Martial came into vogue. England's leading sires Hyperion and Nearco sired their share of two year old winners yet had the capacity to sire classic winners as well; whereas in North America, the distinctly *Arabian type* was transmitted by exceptional sires like Mahmoud and his close relative Nasrullah.

Preakness Stakes winner Polynesian (Unbreakable-Black Polly by Polymelian), a fast horse which set track records, became a nice source of speed for commercial breeders. He sired horse-of-the-year Native Dancer. No better source of soundness could be found than Princequillo (Prince Rose-Cosquilla by Papyrus), a male line descendant of St. Simon with legs of steel. In France, French Derby winner Mieuxce established a powerful influence, but perhaps not as widespread as Tourbillon's champion son Djebel. Ironically, a handicapper named Wild Risk, grandson of Rabelais, was to exert surprising improvement on the breed in France even though his influence came from a humble beginning. Wild Risk (sire of Le Fabuleux) is the duplicated ancestor in the pedigrees of champions Nashwan and Arazi, and traces in female line to Lady Cynosure, full sister to leading sire Polymelus (sire of Phalaris, the sire of Pharos).

Of the mares in Phase 6, Myrtlewood established a strong impact in North America, her descendants including champion Seattle Slew and leading sire Mr. Prospector. Rough Shod II became the ancestress of Sadler's Wells, Nureyev and Apalachee. Lavendula II (by Pharos) produced French champion Ambiorix as well as two celebrated daughters in Source Sucree (dam of champion Turn-To) and Perfume II (dam of champion My Babu and of Sayani). Flaming Top became a grand producer for Windfields Stud in Canada, her descendants being champions Nijinsky II and The Minstrel. Feola is the ancestress of champions Nashwan, Round Table, Aureole, as well as Unfuwain. During Phase 6, the influence of Lady Josephine's descendants played a dramatic role in upgrading speed.

SCAPA FLOW (Chaucer-Anchora) dam of Pharos and Fairway.

Phase 7, incorporating stallions and mares born between 1945 and 1960, did not provide as much improvement as was expected. By now, improvement of speed had reached a plateau and was nearing the end of an exponential curve. Other factors which precipitated a slowing-down of the rate of improvement was the export from Europe of high-class breeding stock to North America, the desire of English and Irish breeders to design matings that were outbreds (thus diluting many influential strains), and without question the tardiness of the English Jockey Club to adapt to changing times. Purses in England fell far short of monies made available for races conducted in North America and France. Too many high-class mares were sold from Europe to other countries after 1945, and it led to a decline in quality which went unrecognized for too long a period by European breeders.

Among the great racehorses listed in the table for Phase 7 mention should be made of unbeaten champion Ribot (twice winner of the Prix de l'Arc de Triomphe), Native Dancer (classic winner of 21 of his 22 races), Tom Fool (horse of the year) and the sensational athlete Swaps, a multiple world record holder. Whereas Ribot and Native Dancer established influential male lines, Swaps (Khaled-Iron Reward by Beau Pere) failed to do so but is today recognized as having sired many important daughters which appear in Grade 1 pedigrees. One of the surprises on the table is Better Self, son of champion Bimelech. He went unrecognized as an important upgrader for classic speed and established his true genetic worth via his female descendants. Tom Fool became the sire of champion Buckpasser, one of the best performers seen in the USA during the 20th century.

Of the mares listed in Phase 7, the pidgeon-toed Somethingroyal (by Princequillo) produced two champion sons, namely Secretariat and Sir Gaylord. Almahmoud (by Mahmoud) established a dynasty and is renowned as the grand-dam of champion Northern Dancer, Arctic Tern and Halo.

Aimee's descendants include champion Blushing Groom. Little Hut produced two great broodmare sires in Habitat and Northfields. Champion filly Grey Flight is another mare whose impact is being felt through linebreedings (the 1996 Preakness Stakes winner is linebred to Grey Flight). Royal Charger's full sister Tessa Gillian established a very strong branch from Mumtaz Mahal's No. 9 family; and Aspidistra (of Tartan Farm fame), is a mare whose reputation is gaining momentum with each passing year. Possibly one or two more high-class stallions and mares can make their way onto the list of Phase 7, before the century comes to its conclusion.

NASRULLAH (Nearco-Mumtaz Begum by Blenheim), leading sire in Europe and North America. He is linebred to Sainfoinand his sister Sierra, and to Oaks winner Canterbury Pilgrim.

Genetic Giants listed for Phase 8 comprise a short period of time, that of only nine years commencing with 1960. This was a remarkable decade in thoroughbred breeding history, its success rate coming as a result of all the clever purchases by North American breeding farms during the 1950's. As a consequence, the strength of the thoroughbred gene pool shifted from Europe to the United States of America.

English, Irish, French and Italian three year old classics were won by many Kentucky-bred colts and fillies. By the 1970's there began a sort of crusade by wealthy Europeans to purchase some of the best colts and fillies at Keeneland's and Saratoga's yearling sales in an attempt to restore the quality of breeding stock. This did not happen overnight, but it was a move in the right direction. Meanwhile, Japan's economy had made a startling recovery by 1960 and Japanese breeders went to Europe to purchase quality stock. Racing in Japan was instituted as far back as 1861, so has a long history. In 1954 we were to learn about the formation of the Japan Racing Association (JRA) whose goal was to improve the thoroughbred racehorse up to an acceptable international standard. Japanese studmasters began successful negotiations to buy English Derby winners and commenced purchasing expensive mares from England's best stakes winning families. Japan's objective was to select breeding stock suited at a mile or more rather than pure sprint lines. Most of Japan's richest races are for middle distance runners. English classic winners which were not receiving full books of mares in Europe were on their short-list.

Champion U.S. two year old Never Bend established an influential extension of Nasrullah's male line by siring colts of the caliber of Mill Reef and Riverman, two champions which carried Princequillo in their dams. Canada's little champion Northern Dancer became the hottest property in the world by siring turf champions like Nijinsky II, Nureyev, Storm Bird, and The Minstrel; Turn-To's male line exerted power via Hail To Reason's English Derby winning son Roberto and via Grade 1 winner Halo; Ribot's sons Graustark, His Majesty and Tom Rolfe maintained a classic line suited for turf or dirt, while French champion Caro and U.S. champion Hoist the Flag rose to prominence on the broodmare sires' list. Whereas Sir Gallahad III and Bull Dog found it difficult to sustain a successful male line of Teddy, horse-of-the-year Damascus (by Sword Dancer out of Kerala by My Babu) was able to do so. Northern Dancer's little son Lyphard became leading sire in France and was soon after purchased for Gainesway Farm, Kentucky. Perhaps the decade of the 1960's in Kentucky was the most fascinating period of the entire 20th century in terms of depth of quality breeding material made available to commercial horse-breeders. Americans had a grip on the best genetic material in the world, and Europeans knew it.

Excellent mares born during the 1960's include English Oaks victress Noblesse (ancestress of Rainbow Quest, Commander in Chief, Warning) , Miss Swapsco (ancestress of champions Devil's Bag, Glorious Song, Rahy), Tamerett (ancestress of Gone West, Lion Cavern, Known Fact, Tentam, Tappiano) while Neriad's female line rose to prominence via the deeds of champion filly Miesque, dam of classic winner Kingmambo. Robert Sangster and Irish trainer Vincent O'Brien teamed with bloodstock agent Tom Cooper to purchase the very best colts sired by Northern Dancer in the mid 1970's and early 1980's. Prices for the stallion's yearlings went to unparalleled heights and Northern Dancer's fee went above $750,000 on a live foal basis. The Keeneland July sales was a mecca for Northern Dancer's fan club. One had to experience the fantasy at first hand to see men's egos in action. One yearling by Northern Dancer fetched more than $13 million, a world record unlikely to be surpassed for some time. In the early 1980's, Sheikh Mohammed bin Rashid al-Maktoum of Dubai (of Bedouin descent) decided to partake a serious interest in horse racing, and armed with a hefty bankroll, went on a spending spree not dissimilar to the Sangster/O'Brien camp. Sheikh Mohammed, who is the crown prince of Dubai and a first-class horseman in every way, influenced his elder brothers Sheikh Maktoum and Sheikh Hamdan to participate in the breeding and racing of thoroughbreds, and as a result, the market for commercial yearlings escalated. Stallion service fees for proven sires in Kentucky rose to record dollar values. It was an unprecedented "boom" market which saw many yearlings bringing in excess of a million dollars a piece.

Phase 9 is still active, thus the list of *Genetic Giants* is incomplete. However, at time of publication it is clear that two young stallions have dynamic genetic impact on the breed to warrant inclusion, namely Sunday Silence and Danehill. The world's best sire during 1995 without question was Breeders' Cup Classic Gr-1 winner Sunday Silence (by Halo) which stands in Japan and dominates his mares to get classic winners. The remarkable Danehill (by Danzig), a horse inbred 3 x 3 to Natalma (the dam of Northern Dancer) was a high-class sprinter-miler in England which found new ownership with Coolmore Stud. Danehill covered mares in three regions of the globe, Ireland, Australia and Japan. Sunday Silence and Danehill have made excellent improvement to the breed and remind me of the same dominance which was transmitted by the great Nasrullah.

In the USA, Mr. Prospector (Raise a Native-Gold Digger by Nashua) must rank as one of the all-time great sires of the 20th Century, despite the fact his success mainly came from dirt runners. An athletic horse with quite small feet, he has created a male line dynasty with high-class sons like Seeking the Gold, Woodman, Miswaki, Forty Niner, Kingmambo, Gone West, Lion Cavern, Machiavellian, Gulch, Afleet, and Fappiano. Another to make his mark on the breed was Blushing Groom (dam by Wild Risk) whose sons include Nashwan, Mt. Livermore, Runaway Groom, Rainbow Quest and Groom Dancer. Daughters of Blushing Groom are making wonderful producers of Group 1/Grade 1 stakes winners. Sons of Northern Dancer to make the list are Danzig, Nureyev, Storm Bird, Sadler's Wells and it is possible Nijinsky II's son Caerleon might be included later. There seems to be a very successful genetic affinity between mixing the strains of Blushing Groom and Northern Dancer. It is possibly the most successful "nick" of the 1990's.

In New Zealand, the Sir Ivor stallion Sir Tristram (tracing in direct female line to Selene) became a sensational sire of classic winners "down-under". In Australia, Star Kingdom's brilliant son Biscay upgraded speed and became an influential broodmare sire. Biscay is linebred to champion Tracery three times. He sired Bletchingly, a leading sire in Australia.

Of the elite mares born since 1970, mention must be made of English co-champion 2YO filly Height of Fashion, dam of four colts which are stallions. Height of Fashion, bred by The Queen, produced champion Nashwan, as well as Unfuwain, Mukddaam, and Alwasmi. Another outstanding producer is Coup de Folie, inbred to Almahmoud. Owned by Maria Niarchos, she is dam of French champions Machiavellian, Coup de Genie, as well as Hydro Calido and Exit to Nowhere. Nodouble's daughter No Class is dam of four champions, namely Sky Classic, Regal Classic, Grey Classic and Classy'n Smart (dam of champion Dance Smartly). Never Knock is dam of Pleasant Tap and Kentucky Derby winner Go For Gin. From an Argentinian family came Grecian Banner, dam of champion filly Personal Ensign (dam of My Flag, Miner's Mark, Our Emblem) as well as successful sire Personal Flag.

By identifying *Genetic Giants* one should aim to duplicate them in future matings for they carry special genetic codes far superior to the best of the rest. Thoroughbred breeders who manipulate matings (by duplicating famous mares and stallions) must have an advantage over their competitors. I foresee the early part of the 21st century revealing many successful thoroughbreds with pedigrees having duplication of the 15.2 hands high stallion Northern Dancer. Great though he was, this bay stallion was a known sire of "bleeders". How this will impact, in the long-term, for soundness is anyone's guess. With so much Northern Dancer penetrating British and Irish pedigrees, perhaps the drug ban may have to be lifted next century to allow for the use of Lasix. Only some descendants of Northern Dancer will carry recessive genes which might cause bleeding from the lungs, so horse-breeders can be selective. In France and the USA, commercial farms have not over-indulged in Northern Dancer's male line and have made available a much wider choice of male lines than is currently found in England or Ireland.

RAINBOW QUEST (Blushing Groom-I Will Follow by Herbager) won the Prix de l'Arc de Triomphe Gr-1 and is a leading sire in England. He traces to English Oaks winner Noblesse. His dam's sire won the French Derby.

SPEED AND STAMINA FACTORS

" In skating on thin ice, our safety is in our speed."
- Ralph Emerson

A brilliant sprinter in say North America, Australia, or Europe can cover 6 furlongs in a time of 1 minute, 8 seconds and some have even bettered this time. The best sprint performers and juveniles in graded stakes often run 6 furlongs between 1-09.0 and 1-10.0 on turf or dirt providing a suitably fast level racing surface is made available. We have come along way since times of 1-12.0 for the same distance were run at the beginning of the century, mainly because there has been a pronounced shift towards speed by buyers, forcing breeders to focus more on precocious speed. Today, a much higher proportion of each foal crop are born with a more upright shoulder angle as a result of efforts made to breed speed to speed. Many sprinters with upright shoulder angle are unable to sustain their speed to a mile distance and visibly weaken after traveling six furlongs. Race clubs in certain regions of the globe have made a point to promote rich sprints at a distance of 6 furlongs, e.g. Australia's Golden Slipper Stakes for two year olds (1200m) on turf and worth $2 million Aussie dollars. The Breeders' Cup Sprint Gr-1 worth $1 million is another. When I first came to the United States and studied the stakes program on offer, I was surprised to find very few Grade 1 or Grade 2 stakes at distances of 6 furlongs. Instead, a sprinter in the USA needed to go at least 7 furlongs to be competitive at the highest level. There are only two big Graded sprints in the USA. By contrast, England, France and Australia have numerous sprints at Group 1 level, and many sprints at Group 2 and 3 level. In the USA, 68% of all races are at 7 furlongs or less.

During the 20th century, Britain was the first source of brilliant sprinters and precocious two year olds. The strains of Phalaris, Sundridge, Gold Bridge, The Boss, Orby, The Tetrarch and Panorama were constant sources of early maturing speed. Yearlings with well-grown muscle in their forearms and gaskins appealed to take to market because new owners of the sport were impatient for quick results from investment. Yearlings bred to race at 2 years were much preferred to those bred for stamina. By the mid-1950's, the importance of staying races like the Ascot Gold Cup, Prix du Cadran, Goodwood Cup, Doncaster Cup and St.Leger, were declining, and the cost of having a horse in training began to spiral.

If one examines the commercial stallion material on offer around the world, one can easily identify an emphasis on standing sprinter-milers rather than stayers. Stallions which fail to sire two year old winners with their first crop have a difficult time regaining their initial reputation, and hardship in maintaining service fees comparable with their opening quotes. First season sires which hit the jackpot with several stakes performing juveniles are literally swamped by commercial breeders, and their insured values escalate rapidly.

If one is breeding for the commercial market, it is absolutely necessary to understand the popularity of speed for juvenile performance. The market for yearlings is no longer just for owners, but for professional pinhookers too, who look to make a profit by selling horses at the now popular two year old-in-training venues. One needs to study the leading sires' list of two year old winners, watch results for successful first-season sires, and consider breeding to any young stallion which was either champion two year old, or runner-up champion two year old of its year. Breeding your mare to a juvenile champion colt is in itself, a good commercial risk if you ever find

yourself hesitant about how to design a future mating. Without exception, yearling buyers will take a risk on a first-season juvenile champion. Among the most precocious strains in North America during the 1990's are Danzig, Storm Bird, Storm Cat, Mr. Prospector, Gone West, Woodman, Crafty Prospector, Nureyev, Carson City, Relaunch, Wavering Monarch, Forty Niner, Valid Appeal and Miswaki. In Australia, Danzig's son Danehill is most popular.

In Europe, sprint lines descending from Grey Sovereign, Northern Dancer, Thatch, Danzig, Warning and Ahonoora play a major role, and young sires like Mujadil, Wolfhound, Arazi, Machiavellian, Vettori, Shalford and Zafonic, should make their mark.

Young sires based in Kentucky with pedigree patterns likely to supply two year old stakes winners include Gilded Time (US champion 2YO colt), Maria's Mon (US champion 2YO colt), Kingmambo (classic-winning son of Mr. Prospector and champion mare Miesque), Dehere and Brocco (US co-champion 2YO colts) and fast juveniles like Mountain Cat, Holy Bull, River Special, and Phone Trick's son Caller I. D. (third best juvenile of his year).

A more difficult task to make leading freshman sire is a colt which wins a Kentucky Derby or Belmont Stakes because its genotype probably suits more for winners at a mile or more (around two turns). Early precociousness may not be present in a classic winner's make-up (genotype). However, research of Preakness Stakes winners indicates they have a good record as sires of juvenile winners; thus a young stallion like Preakness Stakes Gr-1 winner Pine Bluff (Danzig-Rowdy Angel by Halo) could be a likely prospect to succeed with his first crop.

Sometimes a Kentucky Derby winner will have shown brilliance at 2 years like that of a sprinter, and in this instance, can make a very successful first-season sire. Champion 3YO colt Unbridled (Fappiano-Gana Facil by Le Fabuleux, son of Wild Risk) won 2 of 6 starts as a juvenile before winning the Kentucky Derby Gr-1, Breeders Cup Classic Gr-1, Florida Derby Gr-1 and finishing runner-up in the Preakness Stakes Gr-1 to Summer Squall. In his initial crop, Unbridled sired the smart colt Unbridled's Song - winner of the Breeders' Cup Juvenile Gr-1, Florida Derby Gr-1, Wood Memorial Stakes Gr-1) - as well as Grindstone, 1996 Kentucky Derby Gr-1 winner, making him a "hot" sire at Gainesway Farm. A Japanese offer of $13 million was refused in June, 1996, for Fappiano's son Unbridled. This young stallion might inherit partial prepotency from his double of elite mare Aspidistra, plus duplication of sprinter-miler Rough'n Tumble.

What physical attributes does a sprinter possess? First and foremost, there is an actual difference between a very high-class two year old and that of an older sprinter. Whereas the former matures quickly, possesses lots of fast-twitch muscle fibers (especially forearms and through its hind-quarters) and has knees closed earlier than most other two year olds, by contrast, a high-class sprinter may not have developed peak form or physical strength until three or four years old.

Certain strains or male lines are renowned for producing quick-maturing progeny. Sons of Northern Dancer are typical of the type of stallion which can sire brilliant juvenile runners, especially Danzig. Generally, these runners are of the smaller type, quick to mature, possessing well developed muscle and have knee plates closed early enough to be conditioned early for the racetrack. Descendants of War Relic's line of In Reality share the same characteristics. In Australia, Star Kingdom (Stardust-Impromptu by Concerto) which was runner-up best English 2YO colt behind Abernant, was exceptional in producing champion two year old colts and fillies. Star Kingdom's progeny were never big, but they matured quickly and needed to be trained lightly, with just enough work to keep them "fresh". They came to hand quickly and this trait was transmitted to the next two generations of Star Kingdom's male line. The same can be said about

Grey Sovereign and his sons in England, as well as Fair Trial's line via Court Martial, the Nasrullah and Royal Charger lines in the USA, and Sir Gaylord's son Habitat in Ireland.

Brilliant two year old fillies usually make successful stakes producers if they can be matched with stallions which have been successful in top company at a mile or more. However, breeding brilliant fillies exclusively to pure sprint stallions might supply upright shoulder angles in the resulting progeny which could end up wasting valuable energy because they don't possess fluent action. An upright shoulder angle usually means they will hit the ground too hard with their forelegs. Researching a history of matings using parents both limited to 6 furlong speed has shown disappointing results. This method eventually produced the Quarter-horse, a brilliant over-muscled, long-backed horse with awesome speed over 2 furlongs (a quarter of a mile), but it isn't the way to upgrade for speed at the highest level of performance.

A brilliant two year old filly needs to be matched with a horse that can supply some degree of stamina. The stallion selected should be as near perfect in shoulder angle (i.e. between 45-48 degrees angle) as possible. More than 60% of thoroughbreds I have seen in the United States have upright shoulder angles, some slight, some very upright. Rarely will these performers make a Graded stakes winner.

Therefore, in an average crop perhaps only 40% of yearlings in contention to earn black type with a stakes victory, but of those, many come from such weak female lines that they are already at a disadvantage to runners bred from stakes-producing families. Only a small percentage of foals from an annual crop win a stakes race during their career.

Even slightly unsound brilliant racehorses with correct shoulder angle are preferred by the author to a sound racehorse with weak upright shoulder angle. I have known many leading trainers who firmly believe the most critical conformation requirement in a potential stakes runner is the perfect shoulder angle. It is the starting point for an athlete. The part played by high-class sprinters in the development of the classic runner is all too evident for me to extrapolate. In all cases, winners of the 2,000 Guineas, 1,000 Guineas, Derby or Oaks require an element of true speed in their make-up, whether it be a parent or grandparent. Breeding stamina to stamina on a continual basis will result in a loss of speed, especially loss of valuable fast-twitch muscle fibers. Stamina to stamina will end up producing a steeplechaser.

The commercial breeder and the private breeder (who breeds to race) must select carefully between stallion and mare to ensure there is the right combination of speed and stamina. A champion racehorse or a Group 1/Grade 1 stakes winner must possess more than one muscle type to make a true athlete. Not only should one be examining conformation (physical type), but also muscle fiber, temperament, soundness, air-flow factors, and bone maturity.

Professor Patrick Cunningham stated in his article published in "Scientific American" the following passage:

" The two most likely limiting factors (that restrict performance improvement of the racehorse) both relate to blood circulation and muscle metabolism. The first one is the blood-borne supply of oxygen to the muscles for the regeneration of adenosine triphosphate (ATP), the molecular fuel that provides energy for muscle contraction. The second factor is the rate at which lactic acid is *cleared* from the muscles. Lactic acid is a waste product of the anaerobic (oxygen-free) breakdown of glycogen, an energy rich carbohydrate that provides a second source of ATP. If lactic acid accumulates, a muscle cannot derive as much energy anaerobically. In every race, a running horse relies on both aerobic and anaerobic muscle contractions. Although the flow of blood to the muscles increases dramatically during exercise, a horse's maximum effort cannot be sustained by aerobic contraction alone. Similarly, the depletion of glycogen reserves in the muscles and liver transmits activity based on anaerobic contraction."

It depends of course on the distance of the race. Some sprinters are bred to get a longer distance but suffer from metabolic upsets, or else are afflicted by weaknesses to their air-flow or suffer spinal injury. For example, an increasing number of thoroughbreds are born with left laryngeal paralysis, flapper problems, or develop what is known as an entrapped epiglottis (thus, yearlings need to be scoped to detect respiratory problems). Others suffer bleeding attacks, be they mild or serious. There are many things that can affect performance. However, I believe certain racehorses are bred purely for speed and thus are entitled to weaken after traveling six furlongs distance. They are performing at the distance to which they were bred (i.e., according to their genotype for limited speed). To determine which strains are purely sprint specialist lines, consult race records of stallions and mares, and their grandparents, noting whether performance came at age two years, or later. It's highly unlikely a 6 furlongs specialist could sire an Ascot Gold Cup or Melbourne Cup winner (but I suppose even this is not impossible). I remember studying the race record of unbeaten English two year old The Tetrarch (Roi Herode-Vahren) and thinking to myself he must represent a speed influence. Yet when I examined his performance at stud I noted he sired versatile runners, some which could get the St.Leger distance of one and three-quarter miles. The Tetrarch never raced beyond his two year old season, so we couldn't estimate his distance limitation. Obviously he was not simply a pure sprinter. Thus, one has to be careful and keep an open mind about individuals which were never tested beyond a mile.

INHERITANCE OF SPEED

In the studies conducted by Professor Cunningham and his team at Trinity College, Dublin, they make comment about the assumption that the entire selection and breeding process in thoroughbreds is founded on the belief that racing performance is inherited. I wish to quote another extract from their paper because it is has merit and is worth repeating:

" Attempts to analyze the genetics of performance in a systematic way have evolved some distinguished names including Charles Darwin and Francis Galton. It is only in recent decades, however, that good estimates of the hereditability of performance based on adequate data have been produced. Towards that end, my colleagues and I have made several major analyses of Timeform data, the most recent of which included the end-of-year records for 31,263 three-year-olds that raced between 1961 and 1985. We have attempted to measure whether groups of half brothers or half sisters have ratings more alike than those of randomly assembled groups."

" Similarly, we have also looked at the extent in which the ratings of parents and their offspring resemble one another more than do those of random pairs of individuals selected from two generations. Our best estimate says that track performance, as measured by the Timeform rating, is about 35 percent heritable. In other words, about 35 percent of all the variation that we observe in track performance is controlled by heritable factors and the remaining 65 percent by other influences, such as training and nutrition."

This confirms what students of pedigree research already knew. Genetic probability factors govern every mating and even if we design a sensational match, chances are that only one foal from four repeat matings is likely to become the outstanding performer we anticipate.

Sometimes we get lucky and breed two outstanding foals from the same mating (a result made possible if linebreedings to multiple superior ancestors appear in the pedigree). Bahram, Windsor Lad, Tulyar, Citation, Tulloch, Secretariat, Spectacular Bid and Dancing Brave were all great champions on the racetrack but none sired a son of anywhere near their own racing merit. Why?

MUSCLE FIBERS

Muscle fibers utilize ATP as an immediate source of energy for the contractile process. Smooth muscle fibers contract more slowly than do skeletal fibers. Skeletal muscle fibers are enveloped and achored to connective tissue, and respond to contraction only when a sufficient stimulus is applied by a nerve ending. The information signal required for contraction results from *electrical charges along the plasma membrane* of a muscle fiber. Skeletal muscle cells or muscle fibers are multi-nucleated and vary in length.

GALLANT FOX (Sir Gallahad III-Marguerite by Celt) was an outstanding racehorse and son of a French 2,000 Guineas Gr-1 winner. His male line and that of his brother Fighting Fox is almost extinct.

Once the muscle fibers have obtained their adult nerve and connective tissue relationships, they *cease to multiply*. With further growth and exercise, individual fibers enlarge (this is called hypertrophy) until the musculature of the adult is formed. The enlargement of muscles as a result of prolonged exercise is due mainly to the hypertrophy of *existing* muscle fibers, not to the addition of new ones.

All vertebrates possess two distinct types of skeletal muscle fibers that contract at different rates. These are termed *Fast-Twitch* and *Slow-Twitch* muscle fibers, but *Intermediate (Medium-twitch)* types exist. Fast-twitch is also commonly referred to as quick-twitch.

Fast-twitch fibers most suitably adapt to high-intensity exercise, which when sustained, is largely anaerobic (not requiring oxygen). Longer term, aerobic types of exercise (requiring oxygen) are best served with *Slow-Twitch* muscle fibers. Muscle fibers, like all cells, use adenosine triphosphate (ATP) as a soucre of energy. Muscle fibers derive energy from ATP much in the same way as an automobile engine derives energy from burning gasoline. Fibers convert the stored energy from ATP into mechanical motion, using the addition of oxygen. The energy of ATP is stored by transferring its terminal phosphate group together with much of the energy contained in its chemical bond *to another compound called Creatin* which forms Creatin Phosphate (CP). When ATP loses its terminal phosphate group it then becomes Adenosine Diphosphate (ADP). It is a transfer and storage equation.

During active work, muscle fibers rely primarily on Glucose as their source of energy. These reactions are aerobic, that is, dependent upon the presence of oxygen. However, it is possible for a muscle fiber to contract for some time in an oxygen-free environment (an anaerobic situation). However, since the anaerobic pathway is inefficient, muscle fibers possess large emergency stores of glycogen (a polysaccharide carbohydrate). During strenuous exercise, glycogen is rapidly converted to *Lactic acid,* thereby leading to a sense of fatigue. The next step is cramping and pain in the skeletal muscles. Lactic acid is transported to the liver and to other muscles where it is broken down aerobically.

Muscle strength is directly related to muscle size. Muscle power is defined as the amount of work achieved by a muscle over a specific time interval. Factors influencing muscle power include not only the strength of the muscle, but also its *velocity of contraction* and its *rate of contraction.*

Next we must consider muscular endurance, a factor which is known to be possessed by pure-bred *Arabian* horses. Endurance is influenced by the quantities of glycogen stored in the muscles at the start of a competitve athletic event. A diet rich in carbohydrates is most desirable because it permits the athlete to store much greater quantities of glycogen compared to mixed or high-fat diets. Stored energy is the key to marathon runners requiring endurance. When an athlete undertakes a strenuous training program, muscle hypertrophy occurs mainly as a result of the increase in diameter of already existing muscle fibers.

There are internal changes that accompany this hypertrophy. The number of myofibrils and mitochondria tends to *increase* in proportion to the degree of hypertrophy. The amount of Creatin Phosphate and ATP may likewise *increase* as much as 20-40%. Additionally, glycogen and fat stores within the enlarged muscle fiber are increased, as are the enzymes necessary for aerobic metabolism. Whereas skeletal muscle fibers can have more than one nucleus, smooth muscle cells possess one nucleus, centrally located. Smooth muscle is widely distributed throughout the body, e.g. walls of veins, arteries and lymphatic trunks, skin. Muscles undergo contraction but if muscles were not attached to bones, they couldn't function.

Hence, as they shorten during contraction, they pull upon bones to which they are attached, and can move bones. The normal adult nervous system provides very precise control for muscle groups. Muscles pull, never push. Contraction is their only operation. Great force can be exerted upon a tendon which in turn transmits this pull to whatever bone it may be inserted into. Tendons therefore act as cables.

One can easily compare human muscle types. At the Olympic Games, the sprinter is a person with big thighs, tons of fast-twitch muscle fibers and carrying muscle bulk. By dramatic contrast, the marathon runner is a thin person (often small) with slow-twitch muscle fibers (long sinuous fibers) that are the opposite of the bulk carried by sprinters.

Speed and the ability to accelerate are requirements for a high-class racehorse, yet they must be acquired in tandem with stamina to be of benefit. In the design of matings for sprinter-milers I often create inbreeding to a classic winner, to hype up the foal's genotype and heat up the nervous system. I prefer to inbreed to sound classic winners via progeny of both sexes, i.e. their sons and daughters since this offers better results and provides an increased chance for favorable traits to be deployed on both sides of the foal's collective chromosomal pairs. If one merely duplicates ancestors via sons only, or daughters only, it probably inhibits or limits full expression of valuable genes. The most common practice employed by commercial breeders is to duplicate a stallion solely via his male offspring, so that a foal has the same ancestor sitting on the male line of both parents. I believe this kind of duplication has real limitations and for the most part, is less successful in breeding good colts *unless* there are supporting ancestors elsewhere in the pedigree.

STAMINA

The number of races programmed for tests of stamina is lessening with each decade. Perhaps it represents an enforced change caused by economic considerations, but one hopes the emphasis away from stamina will not have a deleterious affect on the breed. It is part of the evolution of the thoroughbred. Whether we like it or not, staying tests are becoming less important to boost betting turnover at racetracks, for the simple reason that race clubs will not provide large enough purses to enable distance races to attract large numbers of starters. This can be overcome by adjustments to racing programs and by increasing purses for staying tests. If the St. Leger in England was worth as much as the Derby, its former status would be restored. Fortunately, some of the more important staying tests have been saved, but many traditional Graded stakes at a mile and a half have been reduced to a mile and a quarter in the United States of America.

When the thoroughbred was in its infancy, heats were conducted over distances of 4 miles or more which was of sheer entertainment value to spectators. Jockeys used clever tactics during these heats to gain advantages over their opponents. The true test of stamina was established with the coming of stakes races such as the Ascot Gold Cup and Prix du Cadran, both run beyond 2 miles and essentially at a fairly fast clip. In Australia, the Melbourne Cup run over 2 miles on turf became the most popular race "down-under". Every first Tuesday in November this handicap staying test would stop the nation for once-in-a-year gamblers to try to pick the winner. The success of the VRC Melbourne Cup Gr-1 worth in excess of $2.5 million as a pure entertainment spectacle is an ideal race to use as an example of what can be achieved by clever promotion.

Racing needs to be entertaining. It competes with other entertainment sports and gambling casinos and can only take a certain slice of the pie. Too many race club directors fail to honor the fact that to be successful in gaining share of market, one has to promote the sport as an entertainment targeted to a wide socio-economic sector of the population. Race clubs need hefty promotional budgets, innovative management, and should learn to market their product more effectively. No better example can be used than the Victorian Racing Club which stage successful programs throughout the year in Australia and promote racing on an on-going basis to the general public. They stage the VRC Melbourne Cup and have learned the value of marketing skills.

For competitve stakes runners to accomplish speed up to a mile and a half, *Slow-Twitch* muscle fibers and clean air-flow are absolute requirements from their genotypes. Among the most successful staying lines in Britain early this century were the lines of Hurry On, Blandford, Gainsborough, and Son-in-Law. Unbeaten Hurry On (Marcovil-Tout Suite by Sainfoin) was a big horse with plenty of bone which sired several English Derby winners including Coronach. His line survived through Precipitation. Blandford sired four Derby winners including Windsor Lad, Bahram and Blenheim (sire of Italy's champion stayer Donatello II, and of Nasrullah's dam Mumtaz Begum). Blenheim's grandson Alycidon was the most influential source of stamina in Britain during the mid part of the century. Alycidon was linebred to a son and daughter of Swynford (sire of Blandford). French breeders developed stout lines from various English sources as well as homebred products like Ksar, twice winner of the Prix de l'Arc de Triomphe and sire of champion Tourbillon, sire of Djebel (winner of the Prix de l'Arc de Triomphe). Among products bred in France which became famous stayers were Tantieme, Relko, Reliance II, and Herbager. The French breeder has always been fond of stamina, and generally speaking, less concerned about commercialism.

Once upon a time winners of the Ascot Gold Cup in England were the most commercial sires to use for the marketing of yearlings. History shows they made leading sires and benefited the breed's improvement. Today, winners of the same race are not held in high esteem. The emphasis towards speed has altered the value of a Gold Cup winning stallion.

Nearco (Pharos-Nogara by Havresac II) and Hyperion (Gainsborough-Selene by Chaucer) both became versatile sires, taking them to the top of the sires' list in England, affecting breeding policies and changing the traditional philosophy held by commercial horse-breeders in Europe. These two stallions not only sired classic and cup winners but were also able to get their share of two year old winners. Nearco sired Dante (English Derby winner) and Sayajirao (English St.Leger winner) as well as sons like Nasrullah (champion 2YO colt), Royal Charger (good sprinter) and Nearctic (Canadian champion). Nearco became the new "Emperor of Stallions" and his grandsons Bold Ruler and Northern Dancer would dominate the sires' list in North America and Europe, respectively.

Lord Derby's small chestnut stallion Hyperion (he stood only 15.2 hands high) was almost as influential as Nearco, and was the sire of versatile performers including Owen Tudor (English Derby), Aureole (champion stayer), Stardust (sire of Star Kingdom and Smokey Eyes), Khaled (sprinter, sire of champion Swaps), Aristophanes (sire of Forli), and Alibhai (sire of Determine and Traffic Judge). Hyperion's speed line of Tudor Minstrel (Owen Tudor-Sansonnet by Sansovino) was eventually to have an affinity with the strain of Nearco's descendant Northern Dancer, matings which duplicated not only Hyperion himself, but also Nearco's sire Pharos. Northern Dancer's father Nearctic was out of Hyperion's daughter Lady Angela. It came as no suprise either, that Northern Dancer's line had a genetic affinity with mares carrying Alibhai's input, since Alibhai (Hyperion-Teresina by Tracery) shares a common genetic cross with Nearctic's dam Lady Angela (Hyperion-Sister Sarah by Abbot's Trace, son of Tracery).

Before closing this chapter, the pedigrees of Nearco and Hyperion are worth some study. They were both linebred to unbeaten St. Simon, and came from powerful female lines. Nearco was bred by Federico Tesio at Dormello Stud, Italy; whereas Hyperion was bred by Lord Derby at his famous farm in England. Nearco and Hyperion showed early maturing speed as juveniles, but were not outstanding in this respect, despite decent two year old form. However, from age three onwards, when physically strong, they were high-class performers and exhibited the very best form. How much impact from St. Simon's influence affecting their genetic codes will never be known, but I imagine it was substantial. Hyperion won the English Derby.

				Cyllene
			Polymelus	Maid Marian
		Phalaris		Sainfoin
			Bromus	Cheery by *St.Simon*
	Pharos			*St.Simon*
			Chaucer	Canterbury Pilgrim
		Scapa Flow		Love Wisely
			Anchora	Eryholme
NEARCO				*St.Simon*
			Rabelais	Satirical
		Havresac II		Ajax
			Hors Concours	Simona by *St.Simon*
	Nogara			Carbine
			Spearmint	Maid of the Mint
		Catnip		The Sailor Prince
			Sibola	Saluda

				Hampton
			Bay Ronald	Black Duchess
		Bayardo		*Galopin*
			Galicia	Isoletta
	Gainsborough			*St.Simon by Galopin*
			St.Frusquin	Isabel
		Rosedrop		Trenton
			Rosaline	Rosalys
HYPERION				*Galopin*
			St.Simon	St.Angela
		Chaucer		Tristan
			Canterbury Pilgrim	*Pilgrimage*
	Selene			Cyllene
			Minoru	Mother Siegel
		Serenissima		Loved One *(ex Pilgrimage)*
			Gondolette	Dongola

In Nearco's pedigree, Havresac II (inbred 2 x 3 to St. Simon and also carrying St. Simon's full sister Angelica, via Ajax's sire Flying Fox) represents speed. Progeny of Havresac II demonstrated the ability to accelerate more than once during a race, and Nearco inherited this trait, passing it on to most of his foals. Pharos was leading sire in France and had been runner-up to Tracery's best son Papyrus in the English Derby. Catnip's sire Spearmint was the winner of the English Derby. Catnip's dam Sibola was the winner of the 1,000 Guineas. Stamina from Spearmint's sire Carbine (winner of the VRC Melbourne Cup with a record carrying weight) may have been another important factor in Nearco's make-up.

Hyperion's sire Gainsborough won the substitute English triple crown during the war years. He was a smallish horse of about 15.3 hands high, typical of Hampton's tribe, with nice symmetry. Gainsborough's dam Rosedrop possessed perfect shoulders and won the English Oaks. Rosedrop can be found in the pedigree of U.S. champion Sword Dancer, sire of champion Damascus. Hyperion's dam Selene, a tough, sound little mare of only 15.2 hands high, was inbred 3 x 4 to dual classic winning filly Pilgrimage (2,000 Guineas, 1,000 Guineas) via a daughter Canterbury Pilgrim (won English Oaks) and a son, Loved One. Please note the way unbeaten St. Simon was duplicated in the pedigrees of Nearco and Hyperion.

Whereas Nearco was linebred to St. Simon via his sons and daughters (plus St. Simon's full sister Angelica and daughters of Galopin), Hyperion was linebred to St. Simon via two sons, namely St. Frusquin and Chaucer, but possessed a balance of Galopin's sons and daughters.

Nearco and Hyperion were two *Genetic Giants* which not only succeeded in transmitting upgraded racing class to their descendants, but also were responsible for changing traditional attitudes held by commercial breeders. No longer was the true stayer going to be patronized as much as he was in previous years. Versatile speed which could be extended to a classic distance was now in vogue in Europe and North America, and stallions with these attributes were held in the very highest esteem.

NORTHERN DANCER (Nearctic-Natalma by Native Dancer).
His sire Nearctic is by Nearco from Lady Angela by Hyperion.
Northern Dancer sired Nijinsky II, Storm Bird, Danzig etc.

In the pedigree of Danzig (Northern Dancer-Pas de Nom by Admiral's Voyage) we find some interesting genetic elements. This medium-sized bay stallion led the North American Sires' List three times, was a brilliant but unsound racehorse, yet transmitted tremendous speed to his foals. He is becoming a sire of sires with sons like Danehill, Chief's Crown, Polish Precedent, Adjudicating (in Japan) and Green Desert. Here is his pedigree:

				Pharos
			Nearco	Nogara (dam by *Spearmint)*
		Nearctic		*Hyperion*
			Lady Angela	Sister Sarah (to *Tracery)*
	Northern Dancer			Polynesian
			Native Dancer	Geisha
		Natalma		Mahmoud
			Almahmoud	Arbitrator
DANZIG (1977)				Fighting Fox-*Sir Gallahad III*
			Crafty Admiral	Admiral's Lady-*to Bull Dog*
		Admiral's Voyage		Olympia (g'son of *Hyperion)*
			Olympia Lou	Louisiana Lou
	Pas de Nom			Fair Trial by *Fairway*
			Petition	Art Paper (to *Tracery)*
		Petitioner		Felstead (to *Spearmint)*
			Steady Aim	Quick Arrow

What makes his pedigree rich in speed and why was he a prepotent sire? First, his dam Pas de Nom carries similar superior ancestors to Northern Dancer, so reinforcement of genetic influence was given an opportunity to be expressed - single strength genes combining both sides to provide double strength genes. In the above pedigree, Danzig is linebred to Hyperion (champion), Tracery (champion), brothers Pharos and Fairway (Group winners), Spearmint (English Derby winner, probably best 3YO colt of his year), and to brother and sister Sainfoin and Sierra. Crafty Admiral is a stakeswinner with *inbreeding* to brothers Sir Gallahad III and Bull Dog which are sons of Plucky Liege, *best daughter of Spearmint* in the stud book. Nearco's second dam Catnip is also by Spearmint, but all these come via "daughters" of Spearmint. To balance a high-class pedigree you need to have a "son" of Spearmint, and of course this is provided by Felstead, sire of Oaks winner Steady Aim. Felstead is by Spion Kop, "son" of Spearmint.

If Danzig didn't possess Felstead in his pedigree, I doubt he would have made a leading sire. This statement may seem outrageous, but I am totally convinced it is Felstead's input which changed the genetic code into a more balanced pattern for a sire. Had Danzig been sound, he could have become a champion sprinter. By collecting genetic influence via *sons and daughters* of a great ancestor, you deploy the genetic influence on both sides of the chromosomes in a way which develops speed. If one collects influence from an ancestor merely through sons, this has the strange effect of increasing stamina at the expense of speed. Why? I don't know the answer. All I can say is that from studies of thousands of pedigrees, building on sons of a male or female ancestor adds stamina not speed. Spearmint is by Carbine which descends in female line from the same source as Polymelus, sire of the brothers Fairway and Pharos in Danzig's make-up. When Spearmint's father Carbine links up with Polymelus, it nearly always increases classic speed. Unbeaten champion Nearco has this combination in his genotype. Danzig's dam reinforced it.

HEREDITABILITY OF HEART SIZE

AND THE "X" FACTOR

"It is a melancholy truth that great men have their poor relations."
- Charles Dickens

" Almost all aspects of life are engineered at the molecular level, and without understanding molecules we can only have a very sketchy understanding of life itself."
- Dr. Francis Crick

Two years ago, I came across a thought provoking article which could affect the way we think about planning thoroughbred matings. After the Keeneland yearling sales had been concluded, I had dinner with a young Japanese veterinarian named Yochitaka Akiyama, who works for Dr. Koichiro Hayata in the north island of Japan. "Yoshi" as he was frequently called by Kentuckians, raised the subject of heart size and heart scores during our conversation and suggested I get hold of an article written by Marianna Haun for the February 28th, 1994, issue of the Thoroughbred Times since this article was titled "The X Factor" and sub-titled "The Large Heart may be transmitted on the X Chromosome."

After reading the article, I became quite curious, not because the subject matter was new, but because it had a different slant on the study about heart size previously published in a little book by Australian professor, Dr. J. D. Steel, and subsequent material published by another Aussie named Dr. Anthony Stewart, of the University of Melbourne. In the February, 1995 issue of The Bloodhorse Review, a monthly publication servicing Australia's breeding industry, I wrote about Marianna Haun's article pointing out the significance of the information she acquired from an interview with Dr. Gus Cothran, a geneticist working at the University of Kentucky's Gluck Equine Research Center, based in Lexington. Dr. Gus Cothram discovered an omission from research conducted by the Australians and argued that recessive factors of the "X" sex-chromosomes should be considered. I now quote from the article:

"Hereditability of heart score has been estimated to be 38 per cent (Stewart said). Dams have been shown to have a greater influence than sires on heart score of their progeny and, sires have a greater influence on their daughters' heart score than on their sons' heart score. Thus the inheritance of heart size in the racehorse may be sex-linked."

Dr. J. D. Steel collaborated with Dr. Anthony Stewart in 1977. Heart scores were measured in 16 thoroughbred stallions, 51 mares and 87 young racehorses (42 colts and 45 fillies). The coefficient statistics from this study indicated a relationship between heart score and heart size of parents and offspring. The Australian researchers discovered on average stallions had 19 per cent influence on their foals, whereas mares had 59 per cent influence. This represented a dramatic difference!

"Sex linkage would affect the regression coefficients because the dam passes an X chromosome to both sexes of progeny whereas the sire *does so only to his daughters* - his sons receiving his Y chromosome," Stewart said.

"Thus, sons only possess one X chromosome which comes from their dam, while daughters have two, one from each parent. The consistency of high heart scores and large hearts in the maternal lines of these mares illustrates the influence of huge-hearted mares on the heart size of their progeny. Mating these and similar mares with stallions having high heart scores, has more consistently produced progeny with large heart scores," he added.

Surprisingly, Dr. Gus Cothran admitted to Marianna Haun that the regression coefficients in the Australian study from sire to daughter and dam to daughter may even be higher, especially if inherited via *recessive genes* because the daughters could have the heart genes and *not express them.* They would be merely "carriers". This suggests the other X chromosome genes were *dominant,* yet they would have a 50:50 chance of passing the genes on to their future sons and daughters. Dr. Cothran pointed out that the study conducted at Melbourne University couldn't measure those mares which were "carriers". Why hadn't I thought of this as a logical explanation?

He added: "Thus, a daughter of a large hearted horse like Secretariat may or may not express the large heart size, but she will have the ability to pass on the gene to her progeny on a 50:50 basis. If that horse were a *double copy mare* (homozygous or one that had a large heart gene on *both X chromosomes*) she would probably express the heart size and pass the gene on to all of her progeny, both male and female. If a characteristic comes on the X chromosome to a SON it is nearly always expressed; but that is not necessarily so in the FEMALE. There it depends on which X is *dominant.* So, if the sex-linked theory is correct, a mare could have the gene and not a large heart; but she could still pass the large heart gene on to her progeny."

Let's make a simple chart to show what this means, but before doing so, please recall that a male thoroughbred possesses XY sex-chromosomes, whereas a female thoroughbred possesses XX. We know the male inherits his Y chromosome from his sire, and his X chromosome from his dam. However, we don't exactly know how a female gets her two X's. Let me try to explain by using an example. Storm Cat's dam Terlingua was one of the best fillies ever sired by champion Secretariat (Bold Ruler-Somethingroyal by Princequillo). We'll use Terlingua as the example.

Terlingua definitely inherited one of her X chromosomes from Secretariat's dam Somethingroyal. We know this to be fact. However, we don't know which X chromosome she inherited from her dam Crimson Saint, another outstanding racemare.

Crimson Saint (Crimson Satan-Bolero Rose by Bolero) had two X's to pass on to her daughter Terlingua, that of Crimson Satan's dam or that of her own dam Bolero Rose. Terlingua can only take one of these from her dam. How do we know which X chromosome was transmitted to Terlingua? Was it from Crimson Satan's dam, or Bolero Rose? I would guess it was from the latter, but it is pure speculation on my part (although I have personally seen Crimson Saint and nearly all her female descendants, having had to learn this to help Storm Cat's initial stud career at Overbrook Farm).

Now we'll take a look at some examples and apply an element of guesswork. The Y chromosome doesn't carry sex-linked genes, so it is immaterial. X chromosomes with average heart scores are depicted as X(h), and those which carry a large heart size are shown as X(H):

<u>Example A</u>

Sire:		Dam:	
SECRETARIAT		**CRIMSON SAINT**	
Y	**X(H)**	**X(H)**	**X(h)**

In Example A, we define Terlingua's dam Crimson Saint as having only one X chromosome carrying the large heart factor (H), the other being an X chromosome with average heart size (h). In this instance, a foal from the match of Secretariat with Crimson Saint has only a 50% chance of inheriting the superior X chromosome carrying the large heart factor (H). We have four sex chromosomes to play with in this mating, i.e. Secretariat's Y and X(H), and Crimson Saint's X(H) and X(h). Here are all four possible results involving the sex-chromosomes:

1) **Secretariat's Y with Crimson Saints's X(H) equals a COLT foal inheriting his dam's superior X(H) or large heart size.**

2) **Secretariat's Y with Crimson Saint's X(h) equals a COLT foal which does not inherit a large heart factor because its dam transmitted X(h).**

3) **Secretariat's X(H) with Crimson Saint's X(H) equals a FILLY foal which should inherit the much sought after *Double Copy* complement of two large heart size chromosomes X(H) plus X(H). Mares with the *Double Copy* could well be among mares named on my list of *Genetic Giants.* For example, mares of the excellence of La Troienne, Plucky Liege, Almahmoud, Selene, Lavendula II, and Height of Fashion.**

4) **Secretariat's X(H) with Crimson Satan's X(h) equals a FILLY foal which inherits two different X's, one representing large heart size X(H) and one representing average heart size X(h). This FILLY foal has a 50% chance of inheriting the large heart size, and 50% chance of inheriting the average heart size. She can be a "carrier", and pass on the X(H) factor to 50% of her foals.**

Now comes the question: Which of the examples above apply to actual foals from this mating? The Secretariat mating with Crimson Saint produced a colt named Pancho Villa, winner of 7 of 18 starts for $596,734 including the Silver Screen Handicap Gr-2, Bay Shore Stakes Gr-2, King's Bishop Stakes, and was 2nd Jerome Handicap Gr-1, Vosburgh Stakes Gr-1, Gotham Stakes Gr-2. He could really fly and was trained by D. Wayne Lukas. His full brother Border Run won 6 races earning $154,300 and was stakes-placed 3rd Hollywood Juvenile Championship Gr-2. Both stand at stud. It is possible that Pancho Villa represents a colt shown above as (1), and Border Run represents a colt shown above as (2), but we'll never know unless a specific study is conducted to verify this guess.

The Secretariat mating with Crimson Saint produced two fillies, one of which died. Terlingua was an outstanding juvenile filly, the best in the west, and ranked second top 2YO of her generation. Trained by D. Wayne Lukas, Terlingua won 7 races earning $423,896 including the Hollywood Juvenile Championship Stakes Gr-2, Hollywood Lassie Stakes Gr-2, and produced the colt Storm Cat (4 wins from 8 starts, $570,610 1st Young America Stakes Gr-1, 2nd beaten a nose by Tasso in the Breeders' Cup Juvenile Gr-1, World Appeal Stakes), the filly Chapel of Dreams (7 wins $643,912 1st Palomar Handicap Gr-2, Wilshire Handicap Gr-2, Golden Poppy Handicap Gr-3), and the filly Lyphard's Dancer (unraced dam of stakes winner C'Mon Lets Dance).

Does Terlingua represent the result of (3) or (4) in the above equation? If she was a *Double Copy* mare as indicated by type (3), all her colts would inherit large hearts; but if she was representative of type (4), only half the number of her colts would receive the X(H) chromosome, the one with the large heart factor. Because Terlingua produced two colts by Alydar which failed to win stakes, I doubt she is a *Double Copy* mare. I do know that one of her colts named Tiajuana (by Slew O'Gold) showed tons of ability and that only injury prevented him from displaying his true ability. Tiajuana may have inherited the large heart size X(H) from Terlingua. He was exported to India. Storm Cat's half-brothers by Alydar probably inherited Y and X(h), but of course I am guessing, and don't know for sure. Tiajuana might make a successful sire in India. He may have inherited an X(H).

We shall now look at two other examples involving matings with the four sets of sex chromosomes:

Example B

Sire:		**Dam:**	
Y	**X(h)**	**X(H)**	**X(h)**
	average	**large**	**average**
	heart size	**heart size**	**heart size**

Combinations produce:

1) **Sire's Y + Dam's X(H) = COLT foal with large heart size factor**
2) **Sire's Y + Dam's X(h) = COLT foal with average heart size factor**
3) **Sire's X(h) + Dam's X(H) = FILLY foal with two different X's, one with average heart size, the other with large heart size factor**
4) **Sire's X(h) + Dam's X(h) = FILLY foal with two average heart size factors**

In the above equation, the filly which is type (4) probably will be far inferior on the racetrack to her full sister which is type (3). The filly which is type (3) will have a 50% chance to inherit a large heart X(H). Type (3) will be a "carrier" for large heart size. Type (4) will not, because she inherits two X's of average heart size, thus can't pass on anything else.

Example C

Sire:		**Dam:** ***Double Copy Mare***	
Y	**X(h)**	**X(H)**	**X(H)**
	average	**large**	**large**
	heart size	**heart size**	**heart size**

Combinations produce:

1) **Sire's Y + Dam's X(H) = COLT foals with large heart size factor**
2) **Sire's Y + Dam's other X(H) = COLT foals with large heart size factor**
3) **Sire's X(h) + Dam's X(H) = FILLY foal with two different X's, one with average heart size, the other with large heart size**
4) **Sire's X(h) + Dam's other X(H) = FILLY foal with two different X's, one with average heart size, the other with large heart size**

The best result from this mating would be to get a colt foal because both types (1) and (2) should inherit large heart size factors (since the Y male chromosome is unimportant). These colts represent an example of sex-linked inheritance, the important X(H) factor being inherited from the colt's dam. Plucky Liege was a great producing mare with two X(H) chromosomes for large hearts, and no matter what sire she was matched with, her colts would inherit large hearts. She produced Bois Roussel (English Derby), Sir Gallahad III (French 2,000 Guineas), Bull Dog, Admiral Drake (Grand Prix de Paris) but she did not produce stakes winning fillies.

Note: If we substitute the above stallion with one possessing Y and X(H) chromomes, and match him with a *Double Copy Mare* carrying two X(H) chromosomes - i.e., in every instance the X's are X(H) for large heart size, then ALL foals (colts and fillies) should inherit large heart size.

Example D

Sire:		Dam:	
Y	X(H)	X(h)	X(h)
	large	average	average
	heart size	heart size	heart size

Combinations produce:

1) Sire's Y + Dam's X(h) = COLT foals with average heart size
2) Sire's Y + Dam's other X(h) = COLT foals with average heart size
3) Sire's X(H) + Dam's X(h) = FILLY foals with two different X's
4) Sire's X(H) + Dam's X(h) = FILLY foal with two different X's

The best result from this mating would be filly foals, because fillies of either 3) or 4) stand a 50% chance of inheriting a large heart. Please notice, the colts in this instance will not inherit large heart size. This kind of mating has a "sex-bias" in favor of fillies for the racetrack. Sex-Linked genes can favor one sex over another. Surely one can now glimpse (from these examples) how genetic probability varies according to the genes found in X chromosomes. From the combinations of the sex-chromosomes, four different answers can result from the same mating. Imagine then, the thousands of probabilities possible from the autosomal chromosomes (the many thousands of genes stored in all other chromosomes). I have just touched the tip of the iceberg of probability factors.

Filly foals from weak female lines can be upgraded for racing performance *providing* they are by a stallion whose dam carries the important X(H) large heart factor. Continued use of stallions with the X(H) factor (that is, stallions which are sons of powerful mares) will create gradual upgrading of performance, and previously weak female lines can begin to produce stakes winners. The reverse is true too. Fillies from strong female lines can be downgraded for racing performance if their sire's dam carries the X(h) average heart factor. Perhaps this answers why female lines rise and fall. There is nothing more disconcerting than to buy a daughter of a Graded stakes winner which proves to inherit below average ability on the racetrack, and worse still, fails to produce a stakes winner.

From these example, one has to wonder about fillies which carry different X's. Hidden recessive genes are transmitted by female "carriers" (those with two different X's). Therefore, full sisters can inherit different X's, and possess different produce records. Stakes producers will transmit their X(H) factor on a more consistent basis than their X(h) factor. A mare which breeds stakes winning colts but fails to breed stakes winning fillies can still transmit one of her X(H) chromosomes to her daughters which become "carriers" of the large heart size factor.

Dr. Gus Cothran suggests that a large heart gene would likely arise as a result of mutation of the normal gene, and would trace back to a single source. This gene could trace back on the X chromosome trail for hundreds of years without changing. Thus, the large heart found in Eclipse (born 1764) could be from the same mutant gene that produced Secretariat's heart in 1970, or the large heart found in Secretariat's maternal grandsire Princequillo. Marianna Haun wrote:

" The Australians trace the heart score back through the pedigree using a zig-zag from sex to sex, starting with the sire's dam to her sire, to his dam and so on. The X chromosome can also be traced on the bottom side through the dam's sire, and through the tail female lines. In searching for a possible single source, following the trail markers of horses who have had proven large hearts such as Secretariat, Sham, Hyperion, all the way back to Eclipse, a common thread was found on the X chromosomal trail."

"Practically every major broodmare sire zig-zagged on the X chromosome back to the legendary broodmare Pocahontas, born in England in 1837. Pocahontas, through one link on her top side through her sire Glencoe, and through numerous links on her bottomside, traced on the X chromosome trail to Eclipse."

"Although from this distance of history it is impossible to know, Pocahontas (from the number of links to Eclipse she has on her bottom side) was probably what Cothran described as a "double copy" mare. Although she never won a race, through her progeny particularly her sons Stockwell, King Tom and Rataplan and her daughter Araucaria, her influence has been strongly felt in the thoroughbred pedigrees of many outstanding racehorses. The large hearted horses in Australasia including the famous Phar Lap follow the X chromosome to Pocahontas and Eclipse." I cannot vouch from personal research of great broodmares sires' pedigrees that the *majority* zig-zag on the X chromosome back to Pocahontas; but I can state that some famous broodmare sires really do trace in a zig-zag manner to the famous mare. Irrespective of aiming to be precise, the concept is indeed interesting. Marianna Haun went on to write:

" Pocahontas is described by historians as a mare of rare vitality. She lived to be 33 and was producing for approximately 20 years. An example given of her extraordinary vitality was the fact she went on racing while in foal, coming in second in a field of nine in a race at Chatham, England in September 7, 1842 when she was more than five months in foal to Camel. In a random tracking through pedigrees, it was found that the greater the success of broodmares and broodmare sires, the greater the number of links to Pocahontas. According to Cothran, this would be compatible with the sex-linked theory because the larger the number of links, the greater the chance of expression on the large heart characteristic. With the female, this produces the greatest chance of being a "double-copy mare."

I wish to thank Dr. Gus Cothran, Professor Steel and Dr. Anthony Stewart for their opinions and research. The article was compulsive reading and I wanted to quote part of it in this Chapter. The Thoroughbred Times, a major weekly racing and breeding magazine, is to be commended for publishing the X Factor article.

Heart size is just one of the components needed in the racing machine we call a stakes performer. There are many other vital factors needed for a superior athlete. The Australian X chromosome theory intrigues Dr. G. Frederick Fregin, one of the nation's leading equine cardiologists and director of the Marion du Pont Scott Equine Medical Center in Leesburg, Virginia. Dr. Fregin is involved in a study following the sex-linked theory in which he is measuring dams and their offspring. He correlates heart score and ultrasound measurements. He is quoted as saying:

"I have a number of combinations where we've looked at the dam and found normal heart score, yet when you look at her offspring she is producing some outstanding runners with large hearts. So, if the mares have got the right X chromosome and they are passing it on, that's what is important. It doesn't matter if she expressed the characteristic herself."

We study individuals in the breed and make assumptions based on what we see, hear and smell. Taking heart scores of a mare only shows what she is generally able to express - it doesn't indicate hidden recessive genes which lie waiting to be exposed in the mare's foals. This helps explain how valuable genes which aid in the production of a stakes winner might "skip a generation". The mare with one of her two X chromosomes carrying the large heart size factor is a "carrier", and while she may never have won a race, she can still transmit to 50% of her foals the large heart size characteristic. Naturally courage must accompany a large heart or big heart score, for without the will-to-win it doesn't mean a lot; but strength of a heart's walls (lining) might be very important.

Now concerning Pocahontas, daughter of Glencoe, she was probably a *Double Copy* mare possessed with two X(H) large heart factor chromosomes. I will pause and give details of this remarkable mare, for she became a roarer, and although placed in good company, winning a heat over 5 furlongs, she is officially classified as a non-winner.

Pocahontas was born in the spring of 1837. She was by the fifth Earl of Jersey's stallion Glencoe, from Marpessa, purchased by Lord Stradbroke for 230 guineas at the 1837 dispersal of King William's bloodstock from Hampton Court. Pocahontas, her filly foal at foot, came with the sale of the mare. Sold as a yearling to the partnership of Messrs. Forth and Greatorex, Pocahontas ran once unplaced in Newmarket's Criterion Stakes at two years. The following season she was conditioned for the English Oaks but was unplaced, and again failed to hit the board in the Goodwood Cup. The fact she was targeted towards these valuable races speaks volumes for her ability. They were indeed severe tests for a maiden filly.

At four years, Pocahontas made three starts. She was unplaced in the Goodwood Cup, the Cesarewitch Handicap and Cambridgeshire Handicap, all three races represented by many starters. As a five year old, Pocahontas won a heat of five furlongs for a fifty pounds plate, but was unplaced in the final. Pocahontas made a roaring sound on the downs at Newmarket when in training, just like her dam Marpessa, and grand-dam Clare (1824). Marpessa produced one other roarer from her number of foals, namely Jeremy Diddler (1839) which won 4 races at three. Marpessa's son Idas won the English 2,000 Guineas and was not afflicted by wind. Marpessa's daughter Boarding School Miss (1841) by Plenipotentiary won 6 of 22 starts and bred 17 live foals. Her grandson Knight of the Garter won 23 races and was runner-up in the English 2,000 Guineas.

Glencoe must have been the source of stamina needed to change Marpessa's line into a classic producing family. Born 1831, this strikingly handsome chestnut horse came from Herod's line of The Flying Dutchman. Glencoe's dam Trampoline was by the influential sire Tramp. She was a half-sister to English Derby winner Middleton.

Trampoline's half-sister Filagree produced Riddlesworth, winner of the English 2,000 Guineas, Charlotte West (winner of the English 1,000 Guineas) and the renowned broodmare Cobweb (winner of the English Oaks and 1,000 Guineas). It was an exceptionally strong family. Glencoe won the English 2,000 Guineas, finished third in the Derby, won the Goodwood Cup by three lengths but was not allowed by his owner Lord Jersey to run in the St. Leger. At four, Glencoe won the Ascot Gold Cup, his sole start that season. After one covering season at stud in England, Glencoe was sold for $3,000 to Colonel James Jackson of Alabama. Glencoe was responsible for a barrage of classic winners in the USA, upgrading many mediocre mares. He headed the sires' list on eight occasions. At age 27, he was transferred to Blue Grass Stud near Georgetown, Kentucky, but died one year later. Here is a sketch of the produce record of his famous daughter, Pocahontas:

Cambaules, colt by Camel - roarer - placed.
Unnamed colt by Muley Moloch - roarer - never ran.
Dolly Varden, filly by Muley Moloch - Winner of a Sweepstakes.
Indiana, filly by Muley Moloch. Unplaced. Dam of Humming Bird, ancestress of Kasbah (won French Oaks), Kizil Kourgan (won French Oaks, French 1,000 Guineas, Grand Prix de Paris and dam of champion Ksar), and Kizil Sou (dam of Kantar). Ksar won the Prix de l'Arc de Triomphe (twice). Kantar also won the Arc de Triomphe.

The property of Mr. Theobald, Pocahontas next produced:
STOCKWELL, colt by The Baron - outstanding stakes winner. Won English 2,000 Guineas, Great Yorkshire Stakes, Goodwood Racing Stakes, English St. Leger (by 10 lengths),

TERLINGUA (Secretariat-Crimson Saint by Crimson Satan) was second best two year old filly of her generation. She produced Storm Cat, sire of classic winners Tabasco Cat and Sardula. Terlingua's sire won the triple crown. Her dam's sire was champion 2YO colt.

Newmarket Stakes, Grand Duke Michael Stakes, The Whip, and was beaten in a photo in the Ascot Gold Cup. Stockwell was a great sire. His sons include St.Albans, Blair Athol, Lord Ronald, The Duke, Lord Lyon, Belladrum and Doncaster.

RATAPLAN, colt by The Baron (thus full brother to Stockwell). Won 43 races from 73 starts, and 16 hands high. Described as a dark chestnut, he was a very tough stayer winning the Epsom Trial Stakes, Chester Grosvenor Stakes, Manchester Tradesman's Cup, Ascot His Majesty's Plate, Doncaster Royal Sweepstakes, Bath Cup, Brighton Cup, and climaxed his career with victory in the Doncaster Cup (by 6 lengths). He was a leading broodmare sire. His sons include Ben Battle (won Irish Derby), Blinkhoolie (sire of the amazing stallion Wisdom).

In the ownership of Mr. Thellusson, Pocahontas next produced:

KING TOM, colt by Harkaway. Winner of two races at 2, Goodwood Maiden Handicap, Brighton Nursery Handicap. Only ran once at three when runner-up to Andover in the English Derby, beaten a neck, with 2,000 Guineas winner Hermit third. He sustained a tendon injury. At four, he won a handicap at Newmarket, and then broke down when contesting the Cesarewitch Handicap. He sired Kingcraft (English Derby), Hippia (English Oaks), Hannah (English 1,000 Guineas, Oaks, St.Leger), Tomato (English 1,000 Guineas), Skylark etc. A leading sire and broodmare sire, King Tom sired the dam of unbeaten St. Simon.

Strood, colt by Chatham. Non-winner. Injured.

In the ownership of Lord Exeter, Pocahontas produced:

AYACANOORA, filly by Birdcatcher (sire of The Baron). Stakes winner of the Hopeful Stakes, Newmarket Column Stakes, she became a useful producer at Hampton Court Stud and bred the colt Chatanooga (Criterion Stakes, sire of Wellingtonia) and the fillies Cestus and Cachuca. Cestus is third dam of Ayr Laddie, Milford, and fourth dam of Belfonds, winner of the French Derby. Cachuca (1860) is grand-dam of Limestone (Prince of Wales's Stakes, Sussex Stakes, Epsom Royal Stakes) and Quicklime (2nd English 2,000 Guineas, Derby).

Knight of Kars, colt by Nutwith. Winner, and sire of The Colonel (Grand National twice).

Heroine of Lucknow, filly by Nutwith. Unraced. Ancestress of Clamart (Grand Prix de Paris) and Clover (French Derby winner by Wellingtonia, thus inbred to Pocahontas)

KNIGHT OF ST. PATRICK, colt by Knight of St. George. 4 wins including Newmarket Column Stakes, Goodwood Bentinck Memorial Stakes. He was a roarer yet sired English 2,000 Guineas winner Moslem, and Solario's fifth dam Beachy Head.

Automation, colt by Ambrose. 3 wins, including Abington Mile (defeating Macaroni). Died young.

AURICULA, filly by Ambrose. Won the Newmarket St. Leger Stakes. Her line has not survived.

Araucaria, filly by Ambrose. Won Stamford Plate (defeating subsequent English 1,000 Guineas winner Siberia). A roarer. She produced for her owner, Mr. Naylor of Hooton Park, the chestnut colt Wellingtonia (4 wins and a leading sire in France). Exported to France, Araucaria produced Apremont (a leading sire in Australia), Camelia (English 1,000 Guineas, English Oaks dead-heated for 1st), Chamant (Middle Park Stakes, Dewhurst Plate, English 2,000 Guineas), Rayon d'Or (English St.Leger by 5 lengths, and an influential sire in USA), and Gardenia (a sister to Camelia which produced Sanctimony, dam of the remarkable French stallion Sans Souci II, won Prix Lupin, Prix Daru and Grand Prix de Paris, and Purity, dam of Rosy Cheeks, grand-dam of classic winners Dante and Sayajirao.

Pocahontas (Glencoe-Marpessa) was purchased at age 30 years by the Marquess of Exeter for ten guineas, and she was not covered. She died at age 33 years, in 1870.

The fact that Pocahontas produced numerous roarers, some of which were decent winners, suggests the noise had nothing to do with unsoundness or left laryngeal paralysis. Many stakes winners were known to be roarers, including classic winner Ormonde. There is no doubt in my mind that Pocahontas was a *Double Copy Mare* whose X chromosomes both carried the large heart factor. Her influence has been spread-eagled throughout some of the best genetic material in the world. Here is the pedigree of the full brothers Stockwell and Rataplan:

				Whalebone
			Sir Hercules	Peri
		Birdcatcher		Bob Booty
			Guiccioli	Flight
	The Baron			*Whisker*
			Economist	Floranthe
		Echidna		Blacklock
			Miss Pratt	Gadabout by *Orville*
STOCKWELL and				Selim
RATAPLAN			Sultan	Bacchante
		Glencoe		Tramp
			Trampoline	*Web*
	Pocahontas			*Orville*
			Muley	Eleanor by *Whiskey*
		Marpessa		Marmion by *Whiskey*
			Clare	Harpalice

Stockwell was a strongly made horse of 16 hands high, and sired Doncaster, the last of his father's English Derby winners and foaled in 1870, the year Stockwell passed away.

In the above pedigree, please note there are three full siblings in the fourth remove (fourth generation of Stockwell and Rataplan) and duplication of Orville three times. The full siblings are WHALEBONE (won English Derby), WHISKER (won English Derby) and their sister WEB, all by Waxy from famous mare Penelope by Trumpator.

Penelope was an outstanding producer. Her other foals include English 1,000 Guineas winner Whizgig (dam of English Oaks winner Oxygen), stakes winner Woful, Web's sister Wire (grand-dam of English Oaks winner Pussy), and Waltz, ancestress of English Derby and St. Leger winner Silvio, Middle Park Plate and Champagne Stakes winner Minting, and English 1,000 Guineas winner Minthe. Maternal power was of the highest class in this female line.

Stockwell's son Doncaster was bred at the Sledmere Stud and purchased for 950 guineas at the Doncaster Yearling Sales by Mr. Robert Peck on behalf of his patron James Merry. Having won the English Derby and being runner-up in the St. Leger, Doncaster was sold to the first Duke of Westminster and in the stallion's first crop came the handsome chestnut colt named Bend Or, founder of a male dynasty whose male descent would from his son Bona Vista, to Cyllene, to Polymelus, to Phalaris, to Pharos, to Nearco (sire of Nasrullah, Royal Charger and Nearctic). Stockwell's sire The Baron (1842) was a dark chestnut horse by Birdcatcher, a horse gifted with black tickings throughout his coat, a recessive gene often passed on to his descendants. The Baron won 5 of 9 starts including the English St. Leger and Cesarewitch, but was a temperamental horse with excessive nervous energy and not easy to train.

Here is the pedigree of the *Double Copy Mare* named POCAHONTAS:

				Woodpecker
			Buzzard	Misfortune
		Selim		Alexander
			Alexander mare	Highflyer mare
	Sultan			Sir Peter
			Williamson's Ditto	Arathusa
		Bacchante		*Mercury*
(Sire)			Mercury mare	Herod mare
Glencoe				Joe Andrews
			Dick Andrews	Highflyer mare
		Tramp		*Gohanna*
			Gohanna mare (*)	Fraxinella
	Trampoline			Pot 8 O's
			Waxy	Maria
		Web		Trumpator
			Penelope	Prunella
				King Fergus
			Beningborough	Herod mare
		Orville		Highflyer
			Evelina	Termagent
	Muley			Saltram
			Whiskey	Calash
		Eleanor		*Diomed*
(Dam)			Young Giantess	Giantess
Marpessa				Saltram
			Whiskey	Calash
		Marmion		*Diomed*
			Young Noisette	Noisette
	Clare			*Mercury*
			Gohanna	Herod mare
		Harpalice (*)		Driver
			Amazon	Fractious

Note: The (*) against the names of the Gohanna mare and Harpalice indicate that these two mares are very closely related.

From the five generation pedigree of Pocahontas, we can see two high-class ancestors were duplicated, namely MERCURY and GOHANNA, and two closely related mares GOHANNA MARE (*) and HARPALICE (*). In the dam's pedigree of Pocahontas, we can see 3 x 3 inbreeding to the influential sire WHISKEY and 4 x 4 to English Derby winner DIOMED.

However, take a closer look, because Muley's dam Eleanor is actually a three-parts genetic sister to Marmion, sire of Clare. ELEANOR and MARMION are both by Whiskey from a daughter of English Derby winner Diomed. This would mean that Marpessa, dam of Pocahontas, could have been a *Double Copy Mare.*

French Derby winner CLOVER (Wellingtonia-Heroine of Lucknow) raced by M. Edmond Blanc. His sire is linebred to two daughters of the Double Copy mare POCAHONTAS.

King Tom (Harkaway-Pocahontas) was an exceptional broodmare sire and his famous daughter St. Angela produced unbeaten St. Simon and Angelica, dam of Orme (sire of Flying Fox and Derby winner Orby). King Tom's daughters were more successful than his sons, both on the racecourse and in the paddocks. In January 1878, King Tom died aged 27 years. In recognition of the contribution King Tom made towards the betterment of the de Rothschild bloodstock program, his deeds were commemorated by the erection of a life size statue, strategically placed in the gardens at Mentmore Stud.

The pedigree of Wellingtonia (Chattanooga-Araucaria by Ambrose) shows 3 x 2 inbreeding to Pocahontas and 3 x 3 inbreeding to Touchstone, a great sire which was only 15.2 hands high. Because Wellingtonia is found in the pedigrees of at least three influential strains he must have been able to transmit large heart size to his descendants. He appears in the pedigrees of:

1) OMNIUM II (Upas-Bluette by Wellingtonia) - French champion KSAR is inbred to Omnium II (3 x 2) and sired French Derby winner TOURBILLON, sire of champion DJEBEL.

2) AJAX (Flying Fox-Amie by Clamart-Alice by Wellingtonia), French champion racehorse which sired TEDDY, sire of ASTERUS, SUN TEDDY, SIR GALLAHAD III, BULL DOG, and the famous producer LA TROIENNE. Ajax was unbeaten winning the French Derby, Grand Prix de Paris, Prix Lupin and Prix Noailles, and proved a wonderful sire of fillies.

3) TRACERY (Rock Sand- Topiary by Orme Plaisanterie by Wellingtonia) - Champion English 3YO colt, winner of the St. Leger, Sussex Stakes, St.James's Palace Stakes, Eclipse Stakes, Champion Stakes and leading sire in Argentina. Tracery sired PAPYRUS (English Derby and broodmare sire of PRINCEQUILLO) and Teresina, dam of leading sire ALIBHAI.

Whenever matings create use of all three sources of WELLINGTONIA (i.e. when the above strains come together in the same pedigree), dramatic improvement of racing class can be expected. I purposely design matings to incorporate genetic impact from Wellingtonia because he is a rare ancestor, carrying two *daughters of Pocahontas*, namely Araucaria and Ayacanoora. Here is the special, clever pedigree of Wellingtonia for us to appreciate:

				Camel
			Touchstone	**Banter**
		Orlando		**Langar**
			Vulture	**Kite**
	Chattanooga			**Sir Hercules**
			Birdcatcher	**Guiccioli**
		Ayacanoora		**Glencoe**
			Pocahontas	**Marpessa**
WELLINGTONIA				**Camel**
			Touchstone	**Banter**
		Ambrose		**Priam**
			Annette	**mare by Don Juan**
	Araucaria			**Sultan**
			Glencoe	**Trampoline**
		Pocahontas		**Muley**
			Marpessa	**Clare**

NIJINSKY II (Northern Dancer-Flaming Page by Bull Page). Europe's Horse of the Year, this Derby winner is sire of champions Caerleon, Sky Classic, Shadeed.

Nijinsky's dam FLAMING PAGE was champion race-mare in Canada. Her daughter Fleur produced English Derby winner The Minstrel. She comes from a female line which supplies plenty of bone and size.

Now let's take a look at the pedigrees of champion racehorses KSAR, AJAX and TRACERY which inherit Wellingtonia's genes:

			Gardefeu
		Chouberski	Campanule
	Bruleur		*Omnium II (dam by Wellingtonia)*
		Basse Terre	Bijou (dam by Doncaster, by *Stockwell*)
KSAR			Upas
		Omnium II	Bluette *by Wellingtonia*
	Kizil Kourgan		Vigilant
		Kasbah	Katia (*tracing to Pocahontas's daughter* Indiana)

Here we find a stallion with three daughters of POCAHONTAS, two coming via Wellingtonia (namely Ayacanoora and Araucaria) and one via Katia (Indiana). Is it any wonder that Ksar became a great sire and that his son Tourbillon helped French horse-breeder Marcel Boussac establish the best racing stable in Europe?

			Ormonde (g.g'son of *Stockwell)*
		Orme	Angelica by *Galopin*
	Flying Fox		*Galopin*
		Vampire	Irony
AJAX			Saumur (by Dollar)
		Clamart	Princess Catherine (*to Pocahontas)*
	Vampire		Wellingtonia (inbred to *Pocahontas)*
		Alice	Asta

Triple crown winner Flying Fox won 9 of 11 starts including the English Derby, 2,000 Guineas, St. Leger, Eclipse Stakes, Princess of Wales's Stakes, Jockey Club Stakes and sired Adam and Ajax. His line is active via Teddy's descendants and via Havresac II (sire line of Ribot) which is the broodmare sire of Nearco. The dam of champion Ajax was linebred to Pocahontas since Clamart traces back to her.

			Springfield
		Sainfoin	Sanda (dam by *Stockwell)*
	Rock Sand		*St. Simon* (dam by *King Tom)*
		Roquebrune	St. Marguerite (dam by *Stockwell)*
TRACERY			Ormonde (g. g'son of *Stockwell)*
		Orme	Angelica (dam by *King Tom)*
	Topiary		Wellingtonia (*inbred to Pocahontas)*
		Plaisanterie	Poetess

British breeders should never have allowed English champion 3YO colt Tracery to leaves their shores. From a few crops in England before export to Argentina, he sired Papyrus (English Derby), The Panther (English 2,000 Guineas), Transvaal (Grand Prix de Paris), Monarch, Teresina (dam of Alibhai), Traverse (dam of Transmute and Traffic) and Flamboyant (Doncaster and Goodwood Cups). He also sired champions in Argentina, where his son Copyright sired the great Congreve. Tracery was inbred to brother and sister St. Simon and Angelica, and carried substantial duplication of the *Double Copy mare* Pocahontas via four of her progeny. Nearco's breeder Federico Tesio admitted that he liked to make use of Tracery in the Dormello Stud matings. Tesio sent Nearco's grand-dam Catnip to Tracery to produce stakes filly Nera di Bicci.

One of the best three year old fillies racing in North America during 1996 is Antespend (Spend a Buck-Auspiciante by Practicante-Auxey by Right of Way-Gevrey by Cardanil). Bred by Jack Kent Cooke's Elmendorf Farm, Antespend has won at time of writing, 8 of her 11 starts for $636,299 including the Santa Anita Oaks Gr-1, Las Virgenes Stakes Gr-1, Honeymoon Handicap Gr-3, Santa Ysabel Stakes (L), Miesque Stakes Gr-3, Green River Stakes and was runner-up in the Fantasy Stakes Gr-2. My friend Bobby Spalding who manages Elmendorf Farm helped raise Antespend (trained by Ron McAnally) and always believed she would become a Grade 1 athlete. He loved the filly from the time she was born.

Antespend's dam Auspiciante (1981) was bred at Haras La Biznaga in Argentina and also was trained by McAnally in California. Her race record shows 8 wins for $427,982 including the Matriarch Invitational Stakes Gr-1, Ramona Handicap Gr-1, Beverley Hills Handicap Gr-2, and in Argentina, won the Clasico Ramon Blaus Gr-2, Clasico Apertura Gr-3 and was 2nd Premio Copa de Plata Gr-1 and 3rd Gran Premio Cotejo de Potrancas-Jorge Atucha Gr-1. Antespend's dam carries many important duplications of champion Tracery.

Claiborne Farm's influential sire Princequillo, by Prince Rose from Cosquilla by Papyrus (best son of Tracery) was second best three year old colt of his year in North America winning the Saratoga Cup in race record time. Only Count Fleet was superior that year. When Princequillo's strain linked up with sons of Nearco (carrying Havresac II's influence via Nearco) it produced champions like Secretariat, Riverman, Sir Gaylord and Mill Reef.

There was a genetic affinity between Princequillo with Nearco and his best sons Nasrullah and Royal Charger. It produced many Group 1/Grade 1 winners. Did this "nick" have anything to do with the inheritance of sex-linked genes via the X chromosomes carrying the large heart factor? Is it worthwhile designing matings with two or three examples of this nick? I notice successful use of the nick has already been used in the pedigree of Grand Lodge (Chief's Crown-La Papagena by Habitat) winner of the Dewhurst Stakes Gr-1, St. James's Palace Stakes Gr-1, and runner-up in the English 2,000 Guineas Gr-1. This nick comes via Secretariat and Habitat's sire Sir Gaylord.

By coincidence (or is it?) Grand Lodge happens to be linebred to Nogara, dam of Nearco and Niccolo dell'Arca, carries influence from Ajax's sire Teddy via Sir Gallahad III and Bull Dog, and has Tracery coming via a doubling of Princequillo, via Petition's dam Art Paper, plus Nearctic's dam Lady Angela. Grand Lodge was rated English champion 2YO colt and was ranked second best 3YO miler on the International Classification. Alan Porter, a pedigree consultant, rightly points out that Chief's Crown has done exceedingly well with mares carrying the strain of Somethingroyal (by Princequillo), dam of Secretariat and Sir Gaylord.

MALE LINE SURVIVAL - FERTILITY AND ITS CYCLIC FUNCTION

"A little body doth often harbour a great soul." - Proverbs

" A small leak will sink a great ship." - Proverbs

The more I study the descent of successful male lines, the more I become confused. Who would have thought the male line of famous French stallion Teddy (Ajax-Rondeau by Bay Ronald) would survive via two weak links to champion three year old Sword Dancer, and thence to his champion son Damascus rather than via male descendants of Sir Gallahad III or his brother Bull Dog? Sir Gallahad III and Bull Dog covered some of the choicest mares in Kentucky. They led the sires' list and their respective sons Gallant Fox and Bull Lea were successful sires. However, their male line disappeared one generation later. Why?

By contrast, in North America Teddy's son Sun Teddy (1933) and his best son Sun Again did not receive anywhere near the quality of mares as did Sir Gallahad III and Bull Dog, yet their sire line has survived via Damascus, sire of Private Account (sire of Private Terms, Personal Flag, Secret Hello), Timeless Moment (sire of Gilded Time), Bailjumper (sire of Skip Trial), Time For A Change (sire of Fly So Free, Technology), Ogygian, Eastern Echo: while in Europe, Teddy's line survived via the moderate performer Aethelstan (1922), sire of Deiri and Deux Pour Cent, sire of champion Tantieme which won Prix de l'Arc de Triomphe Gr-1, and his son Tanerko.

The Y sex chromosome causing maleness in thoroughbreds does not carry other genes. When viewed under an electro-microscope, the Y chromosome is fairly insignificant when compared with the bigger X chromosome of the female. The history of thoroughbred breeding has always focused more upon colts rather than fillies, a practice quite common during the 18th and 19th centuries. The General Stud Books show many unnamed fillies and mares, yet nearly all colts born received a registered name. Man certainly revered the male species and under-estimated the female species in centuries past.

We are told that all modern stallions trace in direct male line to the three foundations stallions *The Darley Arabian, The Byerley Turk and The Godolphin Arabian.* Rather than take up unnecessary space by listing a series of direct male lines from these three foundation stallions, I prefer to concentrate on the modern male lines, highlighting specific weak links in the chain of descent (see sire line charts later in this chapter).

If male dominance really exists, a successful stallion would have a huge number of sons to carry on his male line, but it doesn't seem to work that way. In fact, the opposite is true. Very few sons of a successful stallion can maintain the male line and it seems to become more difficult with each generation. Nearco, one of the greatest sires this century, had more than 54 sons at stud in all parts of the world, yet today his sire line is carried on by only four sons, namely Nasrullah, Royal Charger, Dante, and Nearctic. If male dominance were significant, there should have been at least 25 of his sons able to maintain the sire line. It never happened. For a sire line to survive, it requires the strength from powerful mares. English Oaks winner Canterbury Pilgrim was able to produce Chaucer (by St. Simon) and Swynford (by John O'Gaunt), thus is responsible for enabling two male lines to survive.

Nogara (Havresac II-Catnip by Spearmint), champion two and three year old filly in Italy, was able to produce Nearco (by Pharos) and Niccolo dell'Arca (by Coronach, son of Hurry On), thus is responsible for enabling two male lines to survive. One wonders why Alydar (Raise a Native-Sweet Tooth by On and On) 14 wins for $957,195 including the Florida Derby Gr-1, Champagne Stakes Gr-1, Travers Stakes Gr-1 which competed against triple crown winner Affirmed, is struggling to keep his male line alive. Of all his Graded stakes winning sons like Alysheba, Easy Goer, Criminal Type, Cacoethes, Talinum, Strike the Gold and Turkoman, we are still waiting for a grandson to make a commercial sire, yet Alydar has 42 active sons at stud.

Even the mighty Secretariat, 16 wins $1,316,608 Horse of the Year at 2 and 3 years and triple crown winner, falls into this same category. He sired colts like General Assembly, Pancho Villa, Risen Star, Tinner's Way, Dactylographer, Kingston Rule, Academy Award and yet we still await a grandson to make a commercial sire.

Spectacular Bid, another Horse of the Year which won 26 of 30 starts to earn $2,781, 608 including the Kentucky Derby Gr-1, Preakness Stakes Gr-1, Marlboro Cup Handicap Gr-1, Florida Derby Gr-1 and Charles H Strub Stakes Gr-1, has yet to sire a colt which might carry on his male line. Why is it so difficult for champion racehorses to maintain a sire line?

In England, one of the superstars of the turf was Gay Crusader, a triple crown winner described as a "smasher", yet he failed to sire a colt good enough to sustain his male line. Unbeaten Blandford colt Bahram, a triple crown winner, sired Persian Gulf and English 2,000 Guineas winner Big Game whose only decent son Makarpura (Big Game-Cap d'Or) was exported to Australia and failed to get a son good enough to maintain the sire line. Bahram's male line depends on Persian Gulf's great-grandsons keeping it alive. Male descendants of sensational European champion Sea Bird II, winner of the Prix de l'Arc de Triomphe Gr-1 and English Derby Gr-1, and American-bred world record setter Swaps, have male lines in danger of disappearing. Swaps sired Laramie Trail, the last of the line. Sea Bird II sired Arctic Tern whose son Bering won the French Derby Gr-1, and he in turn sired 1995 English 2,000 Guineas Gr-1 winner Pennekamp. The above champions were unbelievably superior on the racetrack. Why are their male lines in trouble?

Ironically, on the other hand, moderate performers have been able to form part of a male line chain which is succeeding. No better examples can I give than the moderate Native Dancer horse Atan, sire of Sharpen Up; the moderate Nasrullah horse Red God, sire of Blushing Groom; the moderate Sunglow, sire of champion Sword Dancer who sired Damascus; and the moderate performer Unbreakable, sire of Polynesian which was the daddy of champion Native Dancer. There seems to be no logical sense as to why certain sire lines survive, and others become extinct. The only consistent observation I've been able to deduce from pedigree research (the study of genotypes) is that of those moderate males which succeed in siring a son capable of keeping a sire line alive most of them are out of mares with tremendous maternal strength.

Northern Dancer's male line dominates the 1980's and 1990's in Europe for turf performance while Mr. Prospector's male line dominates the 1980's and 1990's in North America for dirt performance. Turn-To's line via Hail To Reason, Best Turn, and Sir Gaylord seems to be making headway, especially via his grandsons Halo (sire of Sunday Silence, Sunny's Halo, Jolie's Halo, Devil's Bag - sire of Twilight Agenda), Sir Tristram (sire of Grosvenor, Military Plume, Zabeel), Roberto (sire of Kris S., Lear Fan, Red Ransom, Silver Hawk) and Cox's Ridge.

Mr. Prospector (Raise a Native-Gold Digger by Nashua) was a brilliant racehorse which never won a Grade 1 stakes. Because he sired progeny with small feet, he was never really popular with European breeders or buyers, despite siring turf runners like Machiavellian (champion 2YO colt in

France), Woodman (champion 2YO colt in Ireland), Kingmambo (French 2,000 Guineas Gr-1, St. James's Palace Stakes Gr-1), Tersa (champion 2YO filly in France), Ravinella (champion 2YO filly in England), Distant View (champion miler in England), Coup de Genie (champion 2YO filly in France), Jade Robbery and Miswaki.

Mr. Prospector was idolized by North American breeders because he was consistent in siring champions during his stud career. Joss Colllins of the British Bloodstock Agency told me he believed Mr. Prospector was one of the greatest sires of the century. I agree with this statement because I have seen most of his best runners and they have the ability to accelerate at a moment's notice. He sired sound racing stock even if many of them inherited slight rotation of their left fore cannon and toed-out in the right foot.

Originally based in Florida before being transferred to Claiborne Farm in Kentucky, Mr. Prospector, son of champion Raise a Native, sired more than 144 stakes winners which earned $64 million including Conquistador Cielo (Horse of the Year), Seeking The Gold, Gulch (champion sprinter). Rhythm (champion 2YO colt), It's in the Air (champion 2YO filly), Eillo (champion sprinter), Queena (champion Older Mare), Gold Beauty (champion sprinter), Afleet, Miner's Mark, Gone West and Fappiano.

He is a successful sire of sires, yet his full brother Search For Gold, a stakes placed winner, has an entirely different sire record. Here is an instance of one brother which displayed speed making a first rate sire, while the moderate performed full brother proves to be just a moderate success. Obviously these full brothers inherited *different portions* of their parental gene pool. Their dosage is given the same numbers, but their *genotype* is quite different.

Of all Northern Dancer's sons which went to stud it was the moderate performer Danzig which three times led the sires' list in North America. A bay horse of 15.3 hands high, Danzig (Northern Dancer-Pas de Nom by Admiral's Voyage) was bred by Derry Meeting Farm and sold as a yearling. A smallish plain horse, he only started three times for 3 wins and $32,400 and never earned black type. He did possess brilliance, but he was not entirely sound in his front ankles. As a sire, he has been a remarkable success for Claiborne Farm's clientele.

Danzig is the sire of 116 stakes winners to date with progeny earnings exceeding $53 million. He is sire of Danehill (leading sire in Australia), Dance Smartly (Canadian Horse of the Year, champion 3YO filly in USA), Chief's Crown (champion 2YO colt, sire of English Derby winner Erhaab), Dayjur (Horse of the Year in England), Lure (champion miler), Polish Precedent (champion 3YO colt in France), Polish Patriot (champion 3YO colt in England), Pine Bluff, Adjudicating, Polish Navy (sire of Kentucky Derby winner Sea Hero), Green Desert, and Maroof. Two of his sons Lure and Maroof experienced fertility problems.

For a stallion to succeed at stud, he must be able to transmit speed and stamina to his progeny on a fairly consistent basis and sire sound runners. If possible, he must come from a strong female line which transmits large hearts and plenty of courage. Most important of all, he must have a powerful broodmare sire to take advantage of any zig-zag, sex-linked genes.

From independent studies conducted in Ireland, Australia, England and the USA, it seems that it is impossible to determine the amount of performance inherited by a stallion's input. We know that heart size is governed by sex-linked factors. From statistics provided for the annual Sires' List we know that very few stallions sire above 12% stakes winners to starters, and those which exceed this per cent are among the great sires we come across so rarely. Usually sires with lifetime records showing a percentage of 12% or more stakes winners ratio to starters, possess "prepotency" - the

ability to breed true for many traits or characteristics. Stallions with some degree of homozygosity caused by a reinforcement of superior ancestors in their ancestry may be capable of transmitting performance to about 33% of their foals and examples of fairly prepotent sires are Northern Dancer, Turn-To, Nijinsky II, Blushing Groom, Halo, Sunday Silence, Mr. Prospector and Danehill.

FERTILITY

Certain male lines exhibit lower or higher fertility than the average, and commercial breeders note these variances. Young stallions with low fertility soon loose their opportunity to cover quality mares because no horse-breeder wants to take too much risk with a valuable stakes winning mare, or stakes producer. To the mare owner, no foal means no income next year.

Kentucky Derby winner Northern Dancer, leading sire in Europe with 146 stakes winners including Nijinsky II, Storm Bird, El Gran Senor, Nureyev, The Minstrel, Secreto, Sadler's Wells, Danzig, Northern Baby, Be My Guest, Dixieland Band, Far North, Lomond, Lyphard, Sovereign Dancer, and Unfuwain, sired numerous sons which proved infertile.

Even among some of Northern Dancer's most successful sons, fertility was below acceptable levels, e.g. El Gran Senor. Even grandsons of Northern Dancer sometimes inherited below acceptable fertility levels, and the latest to make this category is Danzig's son Maroof which was syndicated in New Zealand but got few mares in foal in his initial covering year. Rather than give up, Patrick Hogan is giving Maroof a chance to make good at Cambridge Stud in New Zealand because Maroof is a particularly nice individual with fast-twitch muscle fiber not unlike his grandsire Northern Dancer.

In my opinion, Northern Dancer's sons Topsider, Lyphard and Storm Bird were stallions one could not always rely on for high fertility. Yet, there were other sons which were extremely fertile, Nijinsky II, Danzig and Unfuwain being good examples. Northern Dancer sired several high-priced yearlings which proved to be infertile when they were sent to stud.

The selection process for a racehorse mating rarely takes into consideration fertility. We do not set a goal to breed colts with high fertility. Instead, we set our sights on breeding a stakes winner - the racetrack being the fundamental objective.

Fertility of stallions is affected by a rather complex hormonal cycle, by physical maturity, stress, health, and by a drug-free racing career. A young stallion may start his stud career with *below average* fertility in the initial covering season and yet might follow it up with *above average* fertility in his second covering season. There are many instances of this kind of scenario. Three year old stallions are susceptible to such "slow" beginnings and might be helped by a hormonal stimulant. From my observation, stallions experience bio-rhythmic cycles of high, medium and low fertility levels during a normal year. Among factors which govern a stallion's cycle of maximum or minimum fertility levels are:

> **heat, altitude, climate, nutrition, health (especially high temperature), age, and other environmental considerations.**

Immature sperm takes many weeks to mature and be of use in the breeding shed, so if a stallion shoots blanks, it may be during a short cyclic period when his spermbank has not been replaced quickly enough by sufficient mature sperm. This is likely to be observed in stallions which are required to cover large numbers of mares in both hemispheres during the twelve months, namely shuttle stallions.

If a stallion is to do dual hemisphere covers, he should be selected on his merits re fertility and virility, and an agreement made on the number of covers per breeding season, rather than the number of mares covered, to protect the stallion's breeding future. Some virile stallions can cover more than 160 mares per annum (e.g. National hunt stallions in Ireland do this and more, without many problems) but others may struggle with an annual book of say 60 mares. One has to consider the individual's ability (libido and fertility) to handle large books of mares.

The quality of stallion semen varies from high cycles to low cycles during a twelve month timespan, and during its active life (just like the quality of a mare's follicle). These cycles are, of course, affected by the stallion's age and the amount of sunlight available to act on the pituitary gland. When sunlight is at its minimum in winter, stallions seem to be at the low end of their cycle. When sunlight increases during spring to summer, their cycle peaks. Stallions standing at stud near the equator have different cycles to those which stand in colder climates. The libido of stallions will vary according to their genetic make-up and the size of their testes. Sterility of course happens for no explicable reason. Split heads and tails of active sperm might be of a congenital nature or caused by a bacterial virus.

Health plays an emphatic role. Stallions exposed to extremely high temperatures caused by illness or viral infection, run the risk of becoming partially infertile or totally infertile, because their testes are dramatically affected by heat. Severe heat kills sperm and can cause permanent damage to the stallion's testes. I would rather not express my views about stallions which might have been abused by drugs during their racing careers because this, in my opinion, is a rare event. Sensible owners of a decent Group or Graded stakes winning horse (with sire potential) will protect their investment against risk of abuse. Trainers of horses with obvious sire potential are usually given as a gift one lifetime breeding right when the stakes winner is retired to stud, so it doesn't make sense for a trainer to use harmful drugs.

When buying stallion shares in a young horse, examine your options re mortality and infertility insurance. Never take risks with a first season sire. Always insure against infertility to protect your investment in first season sires. Currently I have a simple formula for estimating the value of a stallion share, irrespective of whether there are bonuses attached to it. This formula is supplied as a guide and works well if the stallion's advertised season price is realistic:

Stallions under 6 years of age: The Advertised season price multiplied by 3.5 = Share value
Stallions 7 to 12 years of age: The Advertised season price multiplied by 3.0 = Share value
Stallions 13 to 17 years of age: The Advertised season price multiplied by 2.5 = Share value
Stallions 18 to 23 years of age: The Advertised season price multiplied by 1.5 = Share value

(Some of my friends use a mutiplying factor of 4.0 for stallions under 6 years of age, but I feel more comfortable with a 3.5 factor).

Therefore, a first season sire aged 5 years of age which is advertised at $10,000 live foal, has a probable share value of $35,000 (maybe $40,000 if x 4.0 is used) based on the above formula; and a proven sire aged 19 years advertised at $30,000 live foal has a probable share value of $45,000.

One has to take into consideration the life expectancy of an older stallion. Also, I make allowance for adding a premium to a share's value should a proven stallion cover a limited number of mares. The number of commercial yearlings sent to market has a definite bearing on the sire's yearling average - a case of supply and demand. Northern Dancer's annual book of mares was restricted to less than 60 mares. This popular commercial stallion had a limited number of colts and fillies made available for public auction (in any given year), thus supply was low when demand was high.

With regard to A.I. (Artificial Insemination) I hope it is never introduced in any Thoroughbred breeding country because it could possibly destroy the economics of the industry. There may be reasons of hygiene and convenience suggested from certain sectors of the thoroughbred breeding industry which seem justified for wanting A.I. introduced, but the downside is so great that I am totally opposed to the idea. I witnessed what happened in New Zealand and Australia in the trotting and pacing industry when A.I. was introduced. Standardbred stallion fees dropped to about one-third of real values and have never made any recovery. Also, A. I. could lessen the number of stallions standing at stud, and therefore less people would be employed. The loss of income from lower stallion fees in New Zealand forced a reduction of spending for new stallions and so the *quality* and *quantity* of imported standardbred stallions (in my opinion) decreased. I looked up the definition of the word artificial, and Webster's dictionary states: "not genuine, made by man artificially".

In any event, does one honestly expect Kentucky, Florida, Newmarket or Kildare farm managements to ever sanction the idea of artificial insemination? Not likely. Their livelihoods would be affected. There are perhaps other ways to improve fertility. The best way is to *select* for fertility when planning matings, and cull mares with congenital infertility problems.

Another important consideration is how a commercial stallion's *Average Price* for yearlings might be affected at public auction. A flood of yearlings by one popular stallion at the *same sales venue* would surely create an over-saturation in that marketplace and depress the stallion's overall average price. Supply would exceed demand. Breeders would likely be exposed to loss of income. Consequently, it is imaginable that lower average prices would put pressure on that sire's advertised fee for the next covering season. No, there are too many reasons why A. I. should never be adopted by the thoroughbred industry and academics which argue its case are not talking through their pockets. Besides, A.I. introduces the chance for dishonest practice and could in the long-term lead to a narrowing of the total gene pool.

In the mid 1980's following the end of a sensational boom period in the blue grass regions of Kentucky, there was a move away from stallion syndication. A number of commercial farms decided to stand stallions and take full risk, thereby ensuring complete *control* of the stallion's annual book of mares. William T. Young was among the first when he decided against syndicating Storm Cat. Seth Hancock, president of Claiborne Farm, also announced that Forty Niner (champion 2YO colt) would stand unsyndicated.

Let me pause for a moment and remind readers about the three main reasons why syndication of stallions was a worthwhile exercise in the 1950's. First, it enabled a farm which wanted to buy an expensive stallion to take in partners, thereby hedging against risk, and raising funds from outside sources. (Claiborne Farm secured Nasrullah by seeking outside investment from friends and clients). Second, it enabled the stallion's book to be filled each year since shareholders in the main owned mares that had to be covered. Third, it enabled stallion shares and seasons to be traded on the open market.

However, what syndication didn't do or couldn't do was to allow the syndicate manager to have "control" over the quality of mares booked, or "control" over share and season prices.

Shares and seasons trading on the open market can sometimes be cheapened by syndicate members wishing to liquefy their assets in a hurry. They might not think about how it will affect other shareholders. The fickle fate of fashion can deal a blow at any time and cause the price of seasons and shares to fluctuate. By standing unsyndicated stallions, "control" was taken back by the farm's management. No outside interference could upset well laid promotional plans.

Stallion fees could be held tight under one source of supply so that bloodstock agents couldn't acquire seasons to trade. Control over the quality of mares accepted to the stallion was once again in the hands of the farm's appointed person. This can be of tremendous value to a stallion manager because it can enable one to seek suitable mares and reject those with unsuitable strains.

After a stallion's progeny appear on the racetrack, computers keep continually updating the results from runners and finally we begin to see how the first season sires or "freshman" sires are faring. After a few crops to race, one can examine the stallion's *Average Earnings* Index (A.E.I) and its *Comparative Index* (C.I) which indicates the quality of mares covered by a stallion.

Taking a recent issue of The Blood-Horse magazine while writing this chapter, the June 1st, 1996 issue showed the following information for Leading Sires of 1996, of which I have taken only a random selection of sires from different male lines to highlight Average Earnings Index and Comparative Index figures:

	CUMULATIVE			
Stallion:	Foals	SW	A.E. Index	Comparative Index
FORTY NINER (by Mr. Prospector)	229	22	3.53	6.15
STORM CAT (by Storm Bird)	319	33	4.46	2.87
DEPUTY MINISTER (by Vice Regent)	452	37	3.17	3.54
THEATRICAL (by Nureyev)	268	26	2.93	3.83
RELAUNCH (by In Reality)	564	74	3.25	2.43
HALO (by Hail to Reason)	678	60	3.24	2.45
MT. LIVERMORE (by Blushing Groom)	334	30	2.90	2.35
WILD AGAIN (by Icecapade)	488	38	2.58	2.27
EASY GOER (by Alydar)	98	3	3.07	6.29
BROAD BRUSH (by Ack Ack)	226	24	3.45	2.38
KRIS S. (by Roberto)	408	36	2.75	1.79
GONE WEST (by Mr.Prospector)	255	32	3,42	3.13
CORMORANT (by His Majesty)	478	40	2.63	1.69
SEATTLE SLEW (by Bold Reasoning)	655	76	4.85	4.98
MR. PROSPECTOR (by Raise a Native)	930	144	4.70	3.41

From the above examples, one can see that some sires can upgrade their mares while others do not if one simply compares their Average Earnings Index with that of the Comparative Index (an index reflecting what the mares have produced to other sires). For example, Storm Cat covered a group of mares producing 319 foals of which 33 were stakes winners, and an Average Earnings Index of 4.46. How does this compare with what those mares produced to other sires? Well, the Comparative Index for those mares shows only 2.87, so Storm Cat has been able to substantially upgrade the progeny of those mares. Let's now look at the deceased sire Easy Goer. He only sired 98 foals of which 3 became stakes winners, giving him an Average Earnings Index of 3.07, but when we look at the Comparative Index, the mares he covered had an index of 6.29 so were obviously top quality producers. Thus, at the present time, I can state that Easy Goer failed to upgrade his mares. On the above list, Kris S. with an Average Earning Index of 2.75 is a clear example of a sire being able to upgrade his mares. The Comparative Index shows only 1.79, so he sure can upgrade for performance. Ribot's grandson Cormorant is another example of a sire which can upgrade his mares.

Sons of small stallions need not inherit the same physical size as their male parent. European champion 2YO and 3YO colt Nijinsky II (Northern Dancer-Flaming Page by Bull Page) was 16.2 hands high compared with the diminutive Northern Dancer at just 15.2 hands high. Nijinsky II is unlike his sire's phenotype and was a much more fertile stallion. It is probable the temperamental Nijinsky II took much of his genotype from certain dominant genes of his dam's sire Bull Page (son of Bull Lea), a rather big Canadian stakes winner with plenty of bone. Flaming Page came from a line of big mares.

Lord Clifden was a reasonably big horse standing 16.1 hands high and taking after his male ancestor Melbourne which was a 16.0 hands high stallion - quite big for those early days when the thoroughbred was finally evolving into the modern racehorse. Lord Clifden became the sire of Hampton which measured only 15.1 1/2 hands high. Hampton sired Bay Ronald which was 15.3 hands high, and he sired Bayardo 16 hands high which sired triple crown winner Gainsborough, a horse of just 15.3 hands high. Whereas Hampton, Bay Ronald and Bayardo were slightly over at the knees, Gainsborough was not. Gainsborough sired English Derby winner Hyperion, which was not quite 15.2 hands high when he won the Epsom classic. Over a forty year timespan I've noticed that the size of a stallion's penis and testes, varies from father to son, and to his sons. It seems nothing wishes to breed true-to-type.

At Mill Ridge Farm in Kentucky, the chestnut stallion Bien Bien holds court. I really like this young stallion because he is so much stronger than his father Manila. Bien Bien is a magnificent horse in the physical sense and won 9 races for $2,331,875 from 2 to 5 years including the Hollywood Turf Cup Stakes Gr-1, San Luis Rey Stakes Gr-1, Hollywood Turf Handicap Gr-1, Swaps Stakes Gr-2, Sunset Handicap Gr-2, San Marcos Handicap Gr-2 and was runner-up in the Santa Anita Handicap Gr-1 and Hollywood Derby Gr-1. He handled turf and dirt and had the speed to compete against sprinters. Even though Bien Bien is son of the small turf champion Manila, earner of nearly $2.7 million, he is a completely different phenotype, stands 16.2 hands high, and has larger sized breeding equipment than his father. No doubt he takes some of his looks from his broodmare sire Graustark.

We now come to examples of successful sire lines, showing some weak links in the chain of descent.

These lists, because of space limitations, are incomplete and meant to indicate only some of the best progeny of successful sire lines. Since I may not have access to information about all active sire lines in every horse-breeding region of the globe (e.g. I find Japan is a difficult country from which to acquire data), I am always fearful that perhaps some forgotten branch of a certain male line will pop up from nowhere and begin to flourish.

Nevertheless, the following charts will give the reader some idea of the major sire lines of the world in 1996:

THE MALE LINE OF TOUCHSTONE
(Eclipse male line)

```
TOUCHSTONE
  Newminster
    Lord Clifden
      Hampton
        Bay Ronald
          Bayardo
            Gainsborough
              HYPERION............................................... |
                Aristophanes                                 High Hat
                | Forli                                      Khaled
                |    Thatch                                    Swaps
                |       Thatching                                 Laramie Trail
                |    Formidable                              Gulf Stream
                |       Efisio                                 Fresh Air
                Alibhai
                | Traffic Judge
                |    Traffic Cop
                |       Deputed Testamony
                Stardust
                | Star Kingdom
                |    Noholme II
                |       Nodouble
                |    Kaoru Star
                |       Luskin Star
                |    Biscay
                |       Bletchingly
                |          Canny Lad
                Aureole
                | Vienna
                |    Vaguely Noble
                |      Exceller
                |  St. Paddy
                |    Jupiter Island
                Owen Tudor
                    Tudor Minstrel
                    | Sing Sing
                    |   Song
                    |   Mummy's Pet
                    |      Aragon
                    | Will Somers
                    |    Balidar
                    |       Young Generation
                    |          Cadeaux Genereux
                    | Tudor Melody
                    |    Welsh Pageant
                    | King of the Tudors
                    Abernant
                      Abwah
```

THE MALE LINE OF STOCKWELL
(Eclipse male line)

STOCKWELL
  Doncaster
    Bend Or
      Bona Vista
        Cyllene
          Polymelus
            Phalaris
              Manna
              | **Colombo**
              | **British Empire**
              | **Endeavour II**
              | **Pretense**
              Pharamond II
              | **Menow**
              | **Tom Fool**
              | **Buckpasser**
              | **L'Enjoleur**
              | **Buckaroo**
              | **Spend a Buck**
              | **Buckfinder**
              | **Track Barron**
              | **Jester**
              | **Reflected Glory**
              | **Tri Jet**
              Fairway
              | **Blue Peter**
              | **Masthead**
              | **Matrice**
              | **Pago Pago**
              | **Island Whirl**
              | **Fair Trial**
              | **Petition**
              | **Petingo**
              | **Pitcairn**
              | **Ela Mana Mou**
              | **March Past**
              | **Queen's Hussar**
              | **Court Martial**
              | **Wilkes**
              | **Vain**
              Pharos
              | **Pharis II**
              | **Pardal**
              | **Pardallo**
              | **NEARCO** **(see separate charts)**
              Sickle
                Unbreakable
                  Polynesian
                    NATIVE DANCER (see seaparate chart)

NEARCO
ROYAL CHARGER
Turn-To
Cyane
| Smarten
Best Turn
| Cox's Ridge
| Sultry Song
Hail To Reason
| ROBERTO
| | Dynaformer
| | Kris S.
| | Brocco
| | Prized
| | Lear Fan
| | Pirate Army
| | Red Ransom
| | Robellino
| | Silver Hawk
| | Hawkster
| Stop the Music
| | Cure The Blues
| | Temperence Hill
| HALO
| Sunday Silence
| Devil's Bag
| Twilight Agenda
| Sunny's Halo
| Dispersal
| Southern Halo
| Strodes Creek
| Jolie's Halo
| Don't Say Halo
| Lively One
SIR GAYLORD
Habitat
| Dalsaan
| Distant Relative
| Lord Gaylord
| Lord Avie
Sir Ivor
Sir Tristram
| Kaapstad
| Kaaptive Edition
| Grosvenor
| Military Plume
| Zabeel
| Octagonal
Bates Motel
Sir Wimborne

HALO (Hail To Reason-Cosmah by Cosmic Bomb), twice leading sire in North America.

NEARCO
NASRULLAH (see separate chart for his champion son BOLD RULER)
Grey Sovereign
| Sovereign Path
| Fortino II
| Caro
| Cozzene
| With Approval
| ZEDDAAN
| Kalamoun
| Kenmare
| Highest Honor
Princely Gift
Fleet Nasrullah
| Gummo
| Flying Paster
Indian Hemp
| T. V. Lark
Nashua
| Good Manners
| Water Bank
Nasram
| Naskra
| Star de Naskra
NEVER BEND
| Mill Reef
| Doyoun
| Shirley Heights
| Darshaan
| Slip Anchor
| Riverman
| Bahri
| Irish River
| Brief Truce
| Exit to Nowhere
Red God
Yellow God
Nebbiolo
BLUSHING GROOM
Nashwan
Arazi
Groom Dancer
Rahy
Mt. Livermore
| Housebuster
| Peaks and Valleys
Rainbow Quest
| Quest for Fame
| Saumarez
Runaway Groom
| Cherokee Run

RIVERMAN (Never Bend-River Lady by Prince John) was a champion in France and a leading sire.

Champion 2YO colt NEVER BEND (Nasrullah-Lalun by Djeddah), sire of Riverman and Mill Reef.

NEARCO
  NASRULLAH
    BOLD RULER
      Blade
      Plum Bold
      Dewan
      Bold Commander
      | Dust Commander
      | Master Derby
      What a Pleasure
      | Foolish Pleasure
      | Marfa
      | Farma Way
      Secretariat
      | Pancho Villa
      | Risen Star
      | General Assembly
      | Academy Award
      Irish Castle
      | Bold Forbes
      | Air Forbes Won
      | Mercedes Won
      | Irish Tower
      | Irish Open
      Chieftain
      | Fit to Fight
      Jacinto
      | Artichoke
      Bold Bidder
      | Cannonade
      | Caveat
      | Spectacular Bid
      Bold Lad (USA)
      Bold Lad (Ire)
      | Persian Bold
      Boldnesian
        Bold Reasoning
        | SEATTLE SLEW
        | Slew O'Gold
        | A. P. Indy
        | Capote
        | Seattle Song
        | Septieme Ciel
        | Slew City Slew
        | Metfield
        | Williamstown
        | Slewpy
        Bold Ruckus
          Beau Genius

NEARCO
  Mossborough
  | Armos
  | Dancing Moss
  | Babor
  Dante
    DARIUS
      Derring-Do
      | Huntercombe
      | Dominion
      | Primo Dominie
      High Top
        Top Ville
          Pistolet Blue

NEARCO
  NEARCTIC
    Icecapade
    | Clever Trick
    | Phone Trick
    | Caller I.D.
    | Tricky Creek
    | Wild Again
    | Wild Harmony
    Briartic
    Explodent
    | Exbourne
    | Explosive Bid
    | Time to Explode
    NORTHERN DANCER----------------------------|
      NIJINSKY II
      | Caerleon
      | Sky Classic
      | Green Dancer
      | Lammtarra
      | Baldski
      | Whiskey Road
      | Strawberry Road
      | Shadeed
      | Alydeed
      Be My Guest
      | Go and Go
      | Media Starguest
      Dixieland Band
      | Dixie Brass
      El Gran Senor
      | Colonel Collins
      Far North
      | The Wicked North
      Lomond
      Local Talent
      Lyphard
      | Dahar
      | Falstaff
      | Lypheor
      | Manila
      | Bien Bien
      Northern Baby
      Northern Park
      NUREYEV
      | Theatrical
      | Robin des Pins
      | Wolfhound
      | Zilzal
      Vice Regent
        Deputy Minister

SADLER'S WELLS
| In The Wings
| Barathea
| Johann Quatz
Secreto
Sovereign Dancer
STORM BIRD
| Storm Cat
| Mountain Cat
| Tabasco Cat
| Catrail
| Future Storm
| Bluebird
| Dolphin Street
| Summer Squall
| Mukaddamah
| Personal Hope
The Minstrel
| Opening Verse
Unfuwain
DANZIG
| Danehill
| Chief's Crown
| Dayjur
| Polish Precedent
| Lure
| Pine Bluff
| Eagle Eyed
| Polish Navy
| Sea Hero
| Green Desert
| Sheikh Albadou
Try My Best
  Last Tycoon
    Marju
    Ezzoud

Polynesian
NATIVE DANCER..|
Atan
| Sharpen Up
| Diesis
| Kris
| Selkirk
| Trempolino
| Sharpo
Jig Time
Restless Native
RAISE A NATIVE
Alydar
| Turkoman
| Alysheba
| Saratoga Six
| Strike The Gold
| Talinum
Exclusive Native
| Affirmed
| Peteski
| Commemorate
| Exclusive Era
| Mill Native
| Premiership
Highland Park
Lines of Power
Majestic Prince
| Majestic Light
| Wavering Monarch
| Maria's Mon
Raise a Man
Raised Socially
MR. PROSPECTOR -------------------------------------|
Fappiano
| Unbridled
| Cahill Road
| Quiet American
| Roy
| Rubiano
Woodman
| Hansel
| Mujtahid
Seeking The Gold
Miswaki
| Black Tie Affair
Afleet
Carson City
Conquistador Cielo
| Mi Cielo

Dan Cupid
Sea Bird II
Arctic Tern
Bering
Pennekamp

French Derby winner BERING, by Arctic Tern.

Crafty Prospector
| Prospector's Gamble
Forty Niner
| Twining
| Marked Tree
Geiger Counter
Gold Stage
Lion Cavern
Gone West
| Zafonic
Gulch
| Thunder Gulch
Jade Hunter
Kingmambo
Machiavellian
| Vettori
Miner's Mark

MALE LINE OF MELBOURNE
(Matchem's male line)

MELBOURNE
  West Australian
    Solon
    | Barcaldine
    | Marco
    | Marcovil
    | HURRY ON
    | Captain Cuttle
    | Coronach
    | Precipitation
    | Summertime
    | Sheshoon
    | Sassafras
    | Saros
    |
    Australian
      Spendthrift
        Hastings
          Fair Play
            MAN O' WAR
              War Admiral
              War Relic
              | Relic
              | Olden Times
              | Full Pocket
              | Fit To Fight
              Intent
                Intentionally
                  Tentam
                  | Baederwood
                  IN REALITY
                    Believe It
                    | Al Mamoon
                    | Believe The Queen
                    Known Fact
                    | Binalong
                    | Warning
                    Kyle's Our Man
                    Relaunch
                    | Skywalker
                    | Bertrando
                    | Waquoit
                    Smile
                    Valid Appeal
                      World Appeal

MALE LINE OF HIMYAR
(Eclipse male line)

```
HIMYAR
  Plaudit
  | King James
  |   Spur
  |   Sting
  |   Questionnaire
  |     Requested
  |     Free For All
  |       ROUGH 'N TUMBLE
  |         Dr. Fager
  |         | Dr. Blum
  |         Minnesota Mac
  |           Great Above
  |             Holy Bull
 Domino
   Commando ............................................ |
     Colin                                              Peter Pan
       Neddie                                             Black Toney
         Good Goods                                       | Black Servant
           Alsab                                          |   Blue Larkspur
             Armageddon                                   |     Blue Swords
               Battle Joined                              Equipoise
                 ACK ACK                                    Swing and Sway
                   Broad Brush
                     Concern
                      Schossberg
```

THE MALE LINE OF SAINFOIN
(mostly found in South America)

```
SAINFOIN
  Rock Sand
    Pilate
      Eight Thirty
    Trap Rock
    TRACERY
      Flamboyant
      Abbot's Trace
      Papyrus
      Copyright
        CONGREVE
          Churrinche
          Corindon
          Hachazo
          Embrujo
            Basajaun
            Penny Post
              Labrador
```

THE MALE LINE OF ROSEVILLE
(Eclipse male line)

```
ROSEVILLE
  Bramble
    Ben Brush
    | Sweep
    |   Eternal
    |     Ariel
    Broomstick
      Whisk Broom II
        John P. Grier
          Boojum
            Snark
              Slide Rule
```

MALE LINE OF ISONOMY
(Eclipse male line)

```
ISONOMY
 Isinglass
  Louvoirs ...........................................................|
  John O'Gaunt                                                    Landgraf
   Swynford...................................|                     Ferro
    BLANDFORD                              St.Germans                 Athanasius
     Bahram                                  The Rhymer                 Ticino
     |  Big Game                               Vertex                     Neckar
     |  | Makarpura                              Lucky Debonair             Zank
     |  Persian Gulf                               Pepenador                Hogarth
     |    Gulf Pearl                                                          Maffei
     |    Tamerlane
     Windsor Lad
     |  Windsor Slipper
     Badruddin
     Blenheim
       MAHMOUD
       |  The Axe II
       |  |  Al Hattab
       |  |  Hatchet Man
       |  |    Naked Sky
       |  Cohoes
      Mirza II
      DONATELLO II
       |  Crepello
       |  |  BUSTED
       |  |     Shernazar
       |  |     Mtoto
       |  |       Shaamit
       |  |     Bustino
       |  |     | Terimon
       |  |     Labus
       |  |       Akarad
       |  |         Tel Quel
       |  Alycidon
       |    Alcide
       |    | Oncidium
       |    Lacydon
       |      Cipayo
       BRANTOME
          Vieux Manoir
            Le Haar
            |  Exbury
            |      Zamazaan
            Vale de Loir
              Tennyson
                 Marasali
```

Robert Sangster, one of the most successful breeders of Group winners this century.

MALE LINE OF DARK RONALD
(mainly found in Germany and Poland)

DARK RONALD
Waldenstein
Prunus
| Oleander
| Orsenigo
Herold
| ALCHIMIST
| | Birkhahn
| | Priamos
| | Literat
| | Surumu
| | Acatenango
| | Gundomar
| | Mangon
| | Baalim
| ARJAMAN
| Olymp
| Masetto
SON -IN-LAW
Beau Pere
| Beau Cheval
| Beau Repaire
| Beaulivre
Foxlaw
| Foxhunter
| Again
| Imbroglio
| Foxbridge
| Fox Myth
Rustom Pasha
| Atabor
BOSWORTH
Plassy
Vandale II
HERBAGER
Sea Hawk II
Matahawk
Big Spruce
Dike
Grey Dawn II
Appiani
Star Appeal
Star Way

TOURBILLON (Ksar-Durban by Durbar II, sire of champion Djebel. He comes from Dollar's male line.

MALE LINE OF DOLLAR
(Herod's male line)

```
DOLLAR
  Androcles
    Cambyses
      Gardefeu
        Chouberski
          Bruleur
            Ksar
              TOURBILLON
                DJEBEL
                | Clarion III
                | | Klairon
                | | | Luthier
                | | |    Twig Moss
                | | |    Tahitian King
                | | |    Saint Cyrien
                | | Lorenzaccio
                | |    AHONOORA
                | |       Dr. Devious
                | |       Topanhoora
                | |       Indian Ridge
                | Le Lavandou
                | | Le Levanstell
                | |    Levmoss
                | My Babu
                |    Better Boy
                |    Crozier
                |       Inverness Drive
                Ambiorix
                | Ambehaving
                | Ambiopoise
                |    Faraway Son
                Hugh Lupus
                | Hethersett
                |    Dalry
                |    Blakeney
                Cardanil II
                Emperor
                Djeddah
                Timor
                   PRONTO
                      Practicante
                      Utopico
                      Tan Pronto
                      Indian Chief
                      Redtop
                      Balconaje
```

MALE LINE OF ST. SIMON
(Eclipse male line)

Vedette
Galopin
ST.SIMON
Florizel II ....|
Doricles
Consols
Massine
Mieuxce
PERSIMMON
| **Prince Palatine**
| **Rose Prince**
| **Prince Bio**
| **Prince Taj**
| **Sicambre**
| **Roi Dagobert**
| **Phaeton**
| **PRINCEQUILLO**
| **Round Table**
| **Apalachee**
| **Flirting Around**
| **Go Marching**
| **Prince John**
| **Silver Screen**
| **Speak John**
| **Hold Your Peace**
| **Meadowlake**
CHAUCER..|
Prince Chimay
Vatout
William of Valence
Vatellor
| **Pearl Diver**
Bois Roussel
Migoli
| **Gallant Man**
| **Gallant Romeo**
| **Elocutionist**
| **Demons Begone**
Tehran
Shikampur
RABELAIS
Biribi
| **Le Pacha**
Rialto
| **Worden II**
| **Wild Risk**
| **Le Fabuleux**
Havresac II
Cavaliere d'Arpino
Bellini
Tenerani
RIBOT
HIS MAJESTY----------------|
Batonnier
Pleasant Colony
Colonial Affair
Pleasant Tap
Cherokee Colony
St. Jovite
Lac Ouimet
Roanoke
Frosty The Snowman
Majesty's Prince
Valiant Nature
Arts and Letters
Tom Rolfe
| **Hoist the Flag**
| **Blue Ensign**
| **Alleged**
| **Muhtarram**
| **Stalwart**
Graustark
| **Key to the Mint**
| **Java Gold**
| **Proud Truth**
Con Brio
Molvedo

MALE LINE OF TEDDY
(Eclipse male line)

```
Ormonde
  Orme
    Flying Fox
      Ajax
        TEDDY
          Sir Gallahad III
          |  Gallant Fox
          |     Omaha
          |     Flares
          |        Chop Chop
          |  Fighting Fox
          |    Crafty Admiral
          |       Admiral's Voyage
          |  Roman
          Bull Dog
          |  Bull Lea
          |     Bull Page
          Asterus
          Ortello
          Aethelstan
          | Deiri
          |    Bey
          |       Beigler Boy
          |          Matun
          |             Mat Boy
          | Deux Pour Cent
          |    Tantieme
          |       Tanerko
          |          Djakao
          Sun Teddy
            Sun Again
              Sunglow
                Sword Dancer
                  DAMASCUS
                    Timeless Moment
                    |  Gilded Time
                    PRIVATE ACCOUNT
                    |  Personal Flag
                    |  Private Terms
                    |  Secret Hello
                    |  Valley Crossing
                    Bailjumper
                    |  Skip Trial
                    Marsyas
                    |  Mister Frisky
                    Ogygian
                    Time For A Change
                      Fly So Free
                      Technology
```

MALE LINE OF SPECULUM
(Eclipse male line, in trouble)

Vedette
  SPECULUM
    Rosebery
      Amphion
        SUNDRIDGE
          Sunstar
            Buchan
          Sunreigh
            Reigh Count
              COUNT FLEET
                Counterpoint
                Beau Gar
                Count Amber
                Auditing
                  Mr. Right
                Count Turf
                Count of Honor

The male line of Roi Herode, sire of unbeaten The Tetrarch, is practically extinct to my knowledge, unless some obscure branch exists in middle Europe or Russia, or a grandson of Alizier comes good in France. For historical purposes, I want Herod's branch of Thormanby to be noted for posterity because it was a speed line which boosted the gray coat color of Le Sancy. Alizier was an outstanding broodmare sire in France. His son Guard's Tie may be the last remaining link.

THE THORMANBY MALE LINE OF THE TETRARCH
(Herod's male line)

Herod
  Woodpecker
    Buzzard
      Castrel
        Pantaloon
          Windhound
            THORMANBY
              Atlantic
                Le Sancy
                  Le Samaritain
                    Roi Herode
                      THE TETRARCH
                        Ethnarch
                        Stefan The Great
                        Salmon Trout
                          King Salmon
                        Tetratema
                          Bacteriophage
                            Teleferique
                              Alizier
                                Guard's Tie

Sundridge's male line has faded, and may not survive. Sundridge was an outstanding speed influence. He sired English Derby winner Sunstar (to which Star Kingdom and Ribot's dam Romanella are inbred) and elite mare Lady Josephine (dam of Mumtaz Mahal, Lady Juror). Sundridge was by Amphion out of Sierra, full sister to English Derby winner Sainfoin.

SAINFOIN is a very special ancestor in pedigrees of Group/Grade 1 pedigrees. He sired triple crown winner Rock Sand (sire of Tracery, and broodmare sire of Man O'War) and his influence was enormous through two special daughters which produced two famous stallions. Sainfoin was the broodmare sire of Phalaris and Hurry On. In my opinion, great sire NASRULLAH owes much of his prepotency to the linebreeding of brother and sister Sainfoin (via Phalaris) and Sierra (via Lady Josephine). Sainfoin belonged to a female line which seemed to transmit large hearts.

SUNDRIDGE (Amphion -Sierra by Springfield-Sanda by Wenlock-Sandal by Stockwell) was a champion sprinter in England. Apart from the fact his dam was full sister to classic winner Sainfoin, Sundridge came from a powerful female line of English Oaks winners Industry and Lady Evelyn, i.e., from the Number 2 Family. His dam Sierra produced the brilliant filly Amphora, winner of the Gimcrack Stakes as a juvenile in England. Amphora established a strong branch in her own right and she was full sister to Sundridge. Amphora is ancestress of champion Cigar, the world's richest racehorse, and ancestress of successful sire Challenger II, an influential broodmare sire which frequently appears in the pedigrees of stakes winners in North America.

CIGAR (Palace Music-Solar Slew by Seattle Slew) has a genetic heritage of interest, so I will provide his direct female line to show its connection with Sainfoin and Sierra. He comes from a branch of Sanda's female line which was developed in Argentina. CIGAR should make a successful sire at Brookside Farm because of his unique pedigree pattern.

Number 2 Family taproot:
Sanda (by Wenlock)
- Sierra (by Springfield) - full sister to English Derby winner SAINFOIN
  - Amphora (by Amphion) - full sister to SUNDRIDGE
    - Flaming Vixen (by Flying Fox)
      - Harpy (by Swynford)
        - Fair Harvest (by Fairway)
          - Vispania (by Tiberius)
            - Agrippine (by Foxhunter)
              - Jungle Queen (by Claro)
                - Gold Sun (by Solazo)
                  - Solar Slew (by Seattle Slew)
                    - CIGAR (by Palace Music)

Palace Music, sire of Cigar, stands in Australia at Segenhoe Stud. In 1996 he was leased to Japan for northern hemisphere cover. Palace Music is a Group winning son of English Derby Gr-1 winner The Minstrel (by Northern Dancer) which is a three-parts brother to Nijinsky II, another English Derby Gr-1 winner. Cigar is raced by his breeder Mr. Allen Paulson, owner of Brookside Farm, at Versailles, Kentucky. After Cigar's victory in the initial $4 million Dubai World Cup, an offer of $20 million was made for Cigar by Japanese interests, but was not accepted. Cigar's dam Solar Slew was sold by Paulson to a breeder in Argentina, but bought back in late 1994 when Cigar's trainer Bill Mott explained to Paulson that he owned a really nice horse. From then on, Cigar began to excite the racing world annexing a string of Graded stakes victories, and taking his sire Palace Music to the top of the sires' list in 1995.

Before closing this chapter, the male line descent from the three famous stallions should be shown.

```
The Darley Arabian
  Childers
     Squirt
        Marske
            ECLIPSE (1764)
               Pot 8 O's
               |  Waxy
               |      Whalebone
               |      |   Camel
               |      |       Touchstone
               |      |          Orlando
               |      |           | Eclipse (USA)
               |      |           |   (then on to Black Toney)
               |      |           Newminster
               |      |             Hampton
               |      |               (then on to Hyperion, Son-in-Law)
               |      Sir Hercules
               |          Birdcatcher
               |          |   The Baron
               |          |       Stockwell
               |          |          Doncaster
               |          |              Bend Or
               |          |                (then on to Nearco, Fair Trial, Northern Dancer)
               |          Oxford
               |              Sterling
               |                 Isonomy
               |                     (then on to Blandford, Crepello)
               |
               King Fergus
                  Hambletonian
                     Whitelock
                         Blacklock
                            Voltaire
                                Voltiguer
                                    Speculum
                                       (then on Sainfoin, Rock Sand)
                                    Galopin
                                       ST. SIMON
                                         (then on to Sicambre, Ribot, Princequillo, Gallant Man)
```

The male lines of the thoroughbred rise and fall according to fashion, opportunities to cover quality mares, and sometimes even their location. St. Simon's sons, and there were many, did not carry on his line in England (his birthplace) and it was left to Rabelais, a son standing in France which covered different strains to St. Simon's pedigree, which provided the strongest male branch. Quantity is never a guarantee that any sire line will survive. In one generation, it can be wiped out.

The Byerley Turk

Jigg
Partner
Tartar
HEROD
Woodpecker
Buzzard
Castrel
| Pantaloon
| Windhound
| THORMANBY
| (then on to The Tetrarch)
Selim
Sultan
Bay Middleton
The Flying Dutchman
Dollar
Androcles
Cambyse
Gardefeu
Chouberski
Bruleur
Ksar
(then on to Djebel, Hethersett, Klairon)

The Godolphin Arabian

Cade
MATCHEM
Conductor
Trumpator
Sorcerer
Comus
Humphrey Clinker
Melbourne
West Australian
| Solon
| Barcaldine
| Marco
| (then on to Hurry On)
Australian
Spendthrift
Hastings
Fair Play
Display
| Discovery
Man O'War
War Relic
(then on to Intentionally, In Reality)

THE MITOCHONDRIA OF CELLS AND ITS MATERNAL IMPORTANCE

" Study the past if you would divine the future."
- Confuscious

The Cub Mare, one of the most important taproot mares in North America, had tremendous influence for speed. A member of the Number 4 Family, she is often quoted in reference material as the "Grandmother of the American Thoroughbred". During 1938 when the Italian-bred colt Nearco raced in France for his owner-breeder Federico Tesio, the unbeaten colt's pedigree was analyzed and showed that he traced directly in female line to this mare.

Imported to New York in 1765 when a three year old, The Cub Mare was owned by Captain James DeLancey, along with the stallion Wildair (GB). Prior to the revolution, the captain dispersed his horses and Wildair (GB) was returned to England but not until The Cub Mare had been matched with him twice. In 1778 a filly foal resulted named Bashaw and then came another called Miss Slammerskin, the latter coming into the ownership of Mr. Hunt of New Jersey. Miss Slammerskin (sometimes referred to as Old Slammerskin) produced the filly named Moll (by Figure), an unnamed filly (by Obscurity) and the filly called Dido (by Bay Richmond). These mares were to establish a racing dynasty.

When a family gets "hot" it constantly produces high-class runners. A genetically powerful female line is capable of succeeding with numerous different strains for some generations. Only with repeatedly poor matings which create unsound conformation, will a stakes performing female line run out of puff. Poor matings may shift the angle of the shoulders, weaken previously good shape of hocks, or maybe lengthen the pasterns, thereby lessening the chance for an athletic foal; but the genes for courage and speed may still exist. Female lines rise and fall according to good or bad selective breedings. Apart from valuable sex-linked genes, there is another bio-chemical element in female thoroughbreds which interests me enough to warrant investigation. I refer to the DNA found in the Mitochondria of cells. When we talk about "luck" in the racing and breeding business, we must always try to steer luck in our direction, and to do this we must try to understand why some horses can out-perform the others.

THE CELL'S ENERGY SOURCE

In choosing a name for this book, I decided to call it "Genetic Heritage" because the starting point for any program is to acquire "quality genes" in breeding stock, or if a non-breeder, "quality genes" in the yearlings to be purchased for the racetrack. My hobby is genetics and in the late 1980's I began to read new material on the importance of the mitochondria of cells. I learned that mitochondria are the source of a living cell's *energy.* Upon further researching the subject, I was amazed to learn that mitochondria possess its own DNA (genetic code) which is inherited strictly from a living organism's mother. Inheritance of mitochondrial DNA is handed down from mother to daughter, and to her daughter, in a direct female chain of descent. I would like to share some of the information I have collected on the subject, because to this day, I am still fascinated by the possible implications of what I now call the "M-factor".

We know that leading commercial farms desire fillies and mares from the best stakes producing female lines. They compete fiercely for them at public auction. Group or Graded stakes winners always command attention, but it is the proven "producer" of Gr-1 or Gr-2 stakes winners which provides the keenest competition of all. It is from these type of mares that tomorrow's best colts might be born, and a really fast colt which wins Gr-1 stakes could be worth more than $10 million upon retirement. One lessens the risk of failure and increaes the chance for success by acquiring proven producers for a breeding program.

When researching the techniques used by Colonel Vuillier and his wife (consultants to the Aga Khan who became leading owner-breeder in Europe), I came across a statement which Madame Vuillier made to an English racing journalist which is worth quoting. She said: "It is always the mare which improves the stallion, and not the stallion which improves the mare."

Federico Tesio made the statement: "The weakest half of any pedigree is usually the bottom half." He referred of course to the dam of a colt or filly.

Wide genetic variation exists in the thoroughbred racehorse. Repeat matings can supply a champion like Mill Reef, or his full brother which struggled to win a lowly Maiden Handicap. Sometimes repeat matings can produce more than one classic winner, e.g. Dante (1st English Derby) and his brother Sayajirao (1st English St.Leger), or a classic winner and a near classic winner like Diminuendo (1st English Oaks) and Pricket (2nd English Oaks). It depends upon whether a special mating can express genetic potential. Breeders of exhibition dogs know that within a litter of six or seven puppies there is sometimes an outstanding puppy capable of being best in show, whereas its siblings may never win a blue ribbon.

Brothers and sisters can be so unlike each other in phenotype because they inherit different "portions" of their parental gene pool. Genetic variation is Nature's way of protecting the survival of the species. We must abide by its functional process and try to understand how genetic probability factors will always upset well laid plans. If there is so much genetic variation between full brothers and sisters, imagine the genetic variation which occurs in a foal bred with a pedigree where no ancestor is duplicated within six generations! Outcross breeding stock transmit enormous genetic variation and possess no prepotency whatsoever.

Thoroughbred breeders have seen with their own eyes the different phenotypes (outward appearance) one mare can produce in her lifetime. Genes express themselves in many forms and some genes which are inherited never get used in the lifespan of an individual. Some genes remain untapped, waiting to be expressed in future generations. Genes are "switched on" and "switched off". For example, when springtime is upon us, mares and stallions shed their winter coats. Genes were "turned on" to grow their long winter coat as protection against a cold environment. By mid-summer, these same animals possess fine coats which gleam in the warm sunlight. Different genes had to be triggered and "turned on" for this to happen. The same applies to the richness of a mare's follicle. The best foals are those which are conceived from the mare's best eggs - her big, ripe follicle rather than her smaller ones. Quality of eggs fluctuates.

I've always admired strong female lines because even when they run out of puff they nearly always return a few generations later with a vengeance. There is something unique about strong female lines and an answer is to be found in their genetic codes (their DNA stored in chromosomes within the nucleus of cells). I try to keep pace with scientific discoveries since my hobby is genetics, so I've explored the basic structure of a cell many times and always search for information about ENERGY when reading articles about genetics. The cells of a racehorse (and human) possess a nucleus and fall into the classification of Eukaryotes - organisms that have cells in which genetic material is located in the nucleus. Within the nucleus are the chromosomes, which carry the genetic code (DNA) for the individual of a species. Outside the nucleus of a cell is *cytoplasm*, a jelly-like substance which contains many important working elements for cell duplication and survival. Within the *cytoplasm* which surrounds the nucleus are "MITOCHONDRIA", the source of energy of cells. Every cell requires energy to do its work and function properly. Energy is of paramount importance to an athlete. For years I suspected a reason why great racehorses inherit superior speed and stamina had something to do with maternal biochemistry. Pedigree research and personal observation indicated success comes via the use of powerful female influence and this opinion was formed after reading many articles and books written by successful horse-breeders and scientists.

The sum total of my gut instincts convinced me that somewhere in the chain of inheritance we were missing out on something important ...... inheritance that was NON-MENDELIAN (not applicable to Mendel's laws of inheritance). ENERGY was a key parameter. Federico Tesio observed nervous energy in the champions he bred at Dormello Stud in northern Italy and made a point about it in his book.

Over the past two decades I grabbed every copy of "Scientific American" which featured articles about genetic discoveries but no article appeared about the mitochondria of cells. Thus I contacted universities, veterinarians and academics to find data about the cell's source of energy. Since the early 1980's, more research focused on the elements within the *cytoplasm* of cells, in particular the mitochondria of cells, When it was learned that the DNA found in the cell's mitochondria can only be inherited via a person's mother, and had different proteins to that of the DNA found in the nucleus of cells, it supplied a medical breakthrough for those searching for treatment of rare diseases that were confined to women.

MITOCHONDRIA OF CELLS

What are mitochondria? Why might they be important? They supply the cell with *energy*. Perhaps some racehorses inherit far *superior* mitochondria from their dams in direct female line descent. This might explain why some families can produce a consistent number of stakes winners. I will now quote from a recently published book called "Fundamentals of Genetics" written by Dr. Peter J. Russell:

"Mitochondria are large organelles which are surrounded by a double membrane, the inner one of which is highly convoluted. Mitochondria play an extremely important role in energy processing for the cell. They also contain genetic material in the form of a circular, double-stranded molecule of DNA. This DNA encodes some of the proteins that function in the mitochondrion, as well as some of the components of the mitochondrial protein synthesis machinery."

"Mitochondria contain their own DNA genomes. The mitochondrial genomes contain genes for the tRNA components of the ribosomes of these organelles, for many (if not all) of the tRNAs are used in organellar protein synthesis and for a few proteins that remain in the organelles and perform functions specific to the organelles. All other proteins are nuclear-encoded, synthesized on cytoplasmic ribosomes and imported into the organelles. At least in the mitochondria of some organisms, the genetic code is different from that found in nuclear protein-coding genes."

The mitochondria of cells have positive and negative affect on the entire biochemistry of an individual (human or racehorse) in terms of energy levels. Geneticists did not fully complete gene maps for the mitochondria of human cells until 1992 even though a fair amount of knowledge was made known about them twenty years ago. Only in the past twelve years has there been substantial, detailed studies focusing upon the mitochondria of cells. This new information, some of it quite surprising, provides geneticists with a new super "biologic" highway.

Molecular biologists investigating the components of living cells discovered that the mitochondria of cells contain their own DNA which is non-Mendelian, and are thus unlike the DNA found in the nucleus of cells. DNA located in the nucleus of a cell contains the genetic code, the chromosomes, the genes, and is labelled *nuclear DNA,* and transmission of inheritance follows the way Gregor Mendel accurately reported from his clever experiments with peas (as described in the Introduction of this book). Nuclear DNA is Mendelian. Geneticists came up with a new description for the DNA found in the mitochondria of cells and called it *mitochondrial DNA,* because it did not follow Mendel's principles of inheritance. Here now, is a layman's interpretation of Mitochondria:

Mitochondria are tiny organelles within the cytoplasm that surrounds the nucleus of living cells. These are present in every living cell and their function is to provide the cell with its energy, which they do by taking the energy from the bonds holding food molecules together, and transferring that energy to so-called high energy phosphate groups which can then be shunted around the cell to various locations. They are tiny sausage-shaped bodies responsible for manufacturing those energy-rich compounds used by the cell. Indeed, they are *power stations* of the cell. Mitochondria reproduce themselves more or less autonomously within the cell, and act as slaves for the cell's well-being. Biologists discovered three surprising things about mitochondria:

1. They possess their own DNA with its slightly different genetic code. *Mitochondrial DNA* is independent, and Non-Mendelian.

2. Mitochondria are passed on to the next generation solely by the mother who hands on some of her mitochondria in the cytoplasm of the egg.

3. The rate of evolution of *Mitochondrial DNA* appears to be approximately ten times *faster* than that of Nuclear DNA (i.e. DNA found in the nucleus of cells).

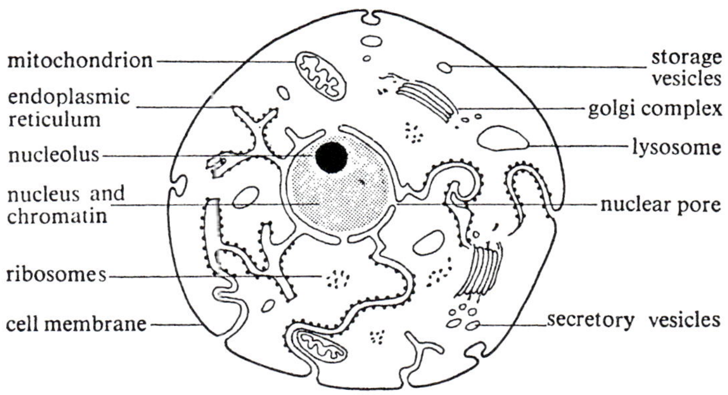

Schematic diagram of a "typical" animal cell.

Mitochondria divide every time a cell divides, regardless of whether sexual reproduction is imminent or not. Each mitochondrion contains a chromosome, a circular strand of DNA, much smaller than a chromosome found in the cell's nucleus.

The mitochondrial chromosome is about 16,000 *bases* long compared to the billions of *bases* of an average nuclear chromosome, yet still manages to contain the codes to produce mitochondrial *enzymes.* All the mitochondria in every cell of the body come from the mother via her egg, and none from the sperm. They are never shuffled around, but can be changed by mutation.

The cell's tiny power plants (mitochondria) allow for most of the reactions of cellular respiration during which fuel molecules are broken down and some of their chemical energy is stored within the energy-rich bonds of ATP (adenosine triphosphate). The energy from ATP can then be used by the cell as needed. Because of its high energy expenditure, a liver cell may contain more than 1,000 mitochondria, and a muscle cell much more.

Each mitochondrion consists of a sack composed of two membranes - the outer one enclosing the contents. Between the membranes is the inter-membrane space, and numerous folds of the *inner* membrane called "cristae" project into the central compartment or matrix. Some of the enzymes needed for cellular respiration are arranged along the "cristae". Other enzymes are dissolved in the contents of the matrix itself. Each mitochondrion has a small amount of DNA and ribosomes can manufacture some of its own proteins, and reproduce themselves.

The nucleus of a cell is a large spherical organelle (see diagram) containing the genetic code. Chemically coded within the nuclear DNA are the instructions for producing all the proteins. Each gene along the chromosomes specifies one protein - one gene equals one protein. Surrounding the nucleus is cytoplasm, and within the cytoplasm are "goodies" that help allow the cell to grow and function, including the mitochondria.

Mitochondria are special because they have their own DNA which is inherited directly from the female parent. In 1994, I received a letter from geneticist Jacinta Flattery (University of New South Wales) in answer to certain requests for information and she replied with a very kind letter together with photostats of recent information about mitochondria. In a post-script at the bottom of her letter to me she wrote: "Are you aware of the function of mitochondria? They are sites at which oxygen is converted to water with the concommitant production of energy. It may be that horses that have "better" mitochondria can perform better...???"

Included in the material she sent to me was data headlined: "Organelle Genes Are Maternally Inherited In Many Organisms." One paragraph contained the following disclosure:

"In higher animals, the egg cell always contributes much more cytoplasm to the zygote than does the sperm, and in some animals the sperm may contribute no cytoplasm at all. One would expect *mitochondrial inheritance* in higher animals to be uniparental or more precisely, *maternal.*"

Geneticists distinguish between the DNA of a nucleus and that of the mitochondria by prefixing the latter as "mt DNA" since it doesn't conform to Mendel's laws. Hence, certain rare diseases found in females are non-Mendelian in the way they are inherited from their mother, a fact which was never considered by medical doctors until biologists confirmed there was another source of DNA in the living cell - the discovery of mitochondrial DNA. Previous medical studies of specific rare diseases in human females were confined to the DNA of the nucleus.

Furthermore, it was discovered when comparing nuclear DNA with "mtDNA" that although the *introns* of the genetic codes were the same, there were differences in the *stop codes* for *exons*, and this really caused a surprise. Studies of *mitochondrial DNA* of three separate races of human females showed a 2% difference in the composition of mtDNA between Anglo-Saxon, Asian, and Aboriginal women. These differences would provide valuable new information for the medical profession.

Here are a few interesting paragraphs from Dr. Francis Crick's book "Life Itself":

"Living organisms must necessarily compete for food, for mates and for all living space, especially with other members of their own species. They must avoid predators and other dangers. For all these various reasons, some will have more offspring than others, and it is the genetic characteristics of such preferred replications which will be passed on preferentially to succeeding generations. In more technical terms, if a gene confers increased "fitness" on its possessor, then such gene is more likely to be found in the gene pool of the next generation. This is the essence of natural selection."

"The first obvious requirement is for replication. Perfect accuracy, on the other hand, is not a requirement. Indeed, all copies should not be exactly the same. Many of the copying errors will be a handicap but a few are likely to be an improvement, allowing the gene to function effectively. We need these for natural selection to operate on; thus, we need mutations, as these genetic errors are called, but not too many of them. There is no use for mistakes which cannot be copied, for these would merely foul up the system. Such mistakes must be eliminated in some way. Confronted with such a chemical error, the copying system may ignore it and put it in one of the standard letters at random. For natural selection to operate, it does not matter all that much what mistake is made as long as the end result is an alteration which can be copied faithfully in the succeeding generations. Replication and mutation are the two essential requirements. In technical terms, an improved gene will usually change not merely the genotype (the collection of genes in an organism) but also its phenotype."

"Protein molecules, being constructed from 20 different side chains, are much more versatile as a class than the nucleic acids DNA and RNA. It is for this reason that all known enzymes are made of proteins, though in certain cases a small organic molecule may be needed to work with it as a co-enzyme. It is the ability of each enzyme to make or break particular chemical bonds which allows modern cells to function at all. Since many different chemical reactions need to be catalyzed in this way, there are many different kinds of enzymes."

AMINO ACIDS

Proteins (genes) comprise specific amino acids. Amino acids are the building blocks of which giant protein molecules are composed. Each amino acid contains nitrogen, carbon, hydrogen and oxygen as well as a variable. There are 20 naturally occurring amino acids in all. Depending on the variable and the pH solution, amino acids may be positively or negatively charged, or neutral, e.g. lysine and arginine have positively charged NH_3+ groups. The ionic proportion of the amino acids mean they can act as buffers in solution. One of the most important chemicals is ATP.

The giant molecules are the stuff of life - our muscles, skin, hair, are protein fibers; beneath the skin, giant molecules of lipid are laid down in layers of subcutaneous fat. Subunits of macromolecules include a number of inorganic ions and up to a thousand or so types of organic molecules, falling into four classes - fatty acids, sugars, amino acids, and the purines and pyrimidines. A number of simple inorganic ions in the cell are positively charged, e.g. potassium, sodium, calcium and magnesium, whilst others are negatively charged, e.g. chloride and phosphate. These ions are crucial for the functioning of the cell, maintaining a livable internal environment. Organic acids - the simple carbon compounds - are acids which take part in complex chemical reactions. Fatty acids are important to the cell's membrane structure. Proteins consist of amino acids and polysaccharides of sugars, whereas nucleic acids contain purines, pyrimidines, sugars and phosphoric acid.

The Genetic Code is derived from the sequential structure of *bases,* using the 20 different amino acids shown in the table on the next page.

THE GENETIC CODE

1st↓ 2nd→	U	C	A	G	↓3rd
U	PHE	SER	TYR	CYS	U
	PHE	SER	TYR	CYS	C
	LEU	SER	—	—	A
	LEU	SER	—	TRP	G
C	LEU	PRO	HIS	ARG	U
	LEU	PRO	HIS	ARG	C
	LEU	PRO	GLUN	ARG	A
	LEU	PRO	GLUN	ARG	G
A	ILEU	THR	ASPN	SER	U
	ILEU	THR	ASPN	SER	C
	ILEU	THR	LYS	ARG	A
	MET	THR	LYS	ARG	G
G	VAL	ALA	ASP	GLY	U
	VAL	ALA	ASP	GLY	C
	VAL	ALA	GLU	GLY	A
	VAL	ALA	GLU	GLY	G

RNA is the nucleic acid which copies the blueprint of the genetic code found in DNA. The flow of genetic information in normal cells runs from DNA to RNA and then to protein, a process considered so fundamentally irreversible by Dr. Francis Crick that he once called it the "central dogma" of molecular biology and is represented thus:

DNA transcription to RNA, then translation to PROTEIN

Incidentally, only a class of viruses called retroviruses (of which HIV, the AIDS-producing virus is but one example) can reverse the flow of DNA to RNA.

RNA is a chain of purine and pyrimidine *bases*, namely Uracil (U), Cytosine (C), Adenine (A) and Guanine (G) which are represented in the above table. Unlike DNA, the sugar component of RNA is not deoxyribose but ribose itself. Also, unlike DNA, its molecules are not composed of two strands inter-twined into a helix, but of single chains. Experiments show that it is on the ribosomes that protein synthesis actually occurs.

Amino acids are primed with sufficient energy for the subsequent formation of peptide bonds and this activation is performed at the expense of ATP by a series of amino acyl synthesase enzymes, one for each amino acid. Amino acids then combine to be incorporated into proteins. The above table shows the relationship between nucleotide triplet "codons" and the amino acids for which they code.

The sixty-four possible combinations of the four RNA *bases* into groups of three is required to code for only twenty amino acids, and there is therefore some redundancy amongst the codons, e.g. three different groups all code for the same amino acid *serine,* namely UCU, UCC, and UCG (see above). Three *bases* = one amino acid. The chemical chain is one of the wonders of inheritance. If 3 *bases* correspond to each amino acid, then 64 (4 x 4 x 4) different amino acids are recognized.

Returning now to the subject of mitochondria in mammals, the transcription of *mitochondrial DNA (mtDNA)* is unusual. Rather than each gene being transcribed independently, both strands of the entire mitochondrial genome are transcribed into two single strands. Mitochondria's codes differ from the nuclear genetic codes. Thus, Dr. Peter Russell quotes:

"Interestingly, the mitochondrial genetic codes are not the same in all organisms. Further differences in the nuclear genetic code have also been found in ciliated promozoans. The genetic code is not as universal as was originally thought."

Then on page 455 of his book, he adds: "Since the pattern of inheritance shown by genes located in organelles differs strikingly from the pattern shown by nuclear genes, the term *Non-Mendelian Inheritance* is used when we are discussing extranuclear genes. In fact, if results are obtained from genetic crosses that do not conform to predictions based on the inheritance of nuclear genes, extranuclear inheritance is a good possibility."

Where is all this information leading us? Perhaps knowledge of *mitochondrial DNA* explains why we have elite mares in the stud books. Although a mitochondrion has its own genetic material, it is not genetically self-sufficient because most of the proteins that account for the structure and function of the organelle (the racehorse) are encoded in the nucleus. The mitochondrion's own DNA specifies the RNA of its ribosomes (on which proteins are assembled), as well as its transfer-RNAs. It helps explain a *direct female transmission of certain unique qualities* via the mitochondria of cells!

THE M-FACTOR

Famous stakes producers probably inherit special *mitochondrial DNA* values from their direct female line, and if coupled with any sex-linked advantages via X chromosomes carrying the genes for large heart size, can become consistent dams of high-class performers on the racetrack. Celebrated mares such as Selene (dam of Hyperion, Sickle, Pharamond II and All Moonshine), Plucky Liege (dam of Sir Gallahad III, Bull Dog, Bois Roussel, Marguerite de Valois), Nogara (dam of Nearco, Niccolo dell'Arca, Nervesa) and Mumtaz Mahal (dam of Badruddin, Mirza II, Mah Mahal and Mumtaz Begum, dam of Nasrullah) are likely to be examples of the type of mares which are elite females, or "super" mums.

Furthermore, I believe the mitochondria of cells of broodmares varies in *QUALITY* so much, that many female lines are unlikely to ever produce Gr-1, Gr-2 or Gr-3 runners. I am referring of course to the bottom 25% of the entire mare population.

These poorer quality mares from inferior damlines are at a distinct disadvantage when compared with the remaining 75% of the mare population; and even if owners of inferior mares use skill in the mating process, these mares are likely to disappoint as producers of above average runners. However, I honestly believe one can upgrade mares in the middle range of the mare population by mating them with compatible stallions that are either sons of Gr-1 winning racemares or else sons of very influential stakes producers. Such stallions as these carry *maternal strength* and can transmit it to the next generation. In a broad genetic sense, stallions are but influential representatives of their dams - their mothers.

It is conceivable that great mares which helped shape the breed (as outlined in the *Genetic Giants* tables) such as Old Bald Peg, Tregonwell's Natural Barb Mare and The Vintner Mare possessed better *mitochondrial* DNA than perhaps all other mares entered in Volume One of the General Stud Book. Their female lines represent numerically, the greater percentage of the overall gene pool, after 200 years of thoroughbred breeding. Incredible, yes?

(Above) NOBLESSE has a facile victory in the English Oaks. She is the ancestress of Rainbow Quest, Warning, andDerby winner Commander in Chief. (Below) The gray mare PERFUME II (Badruddin-Lavendula II by Pharos) dam of English champion 2YO colt and 2,000 Guineas winner My Babu.

Their genetic codes have become widespread in pedigrees of the modern racehorse, and their influence via non-Mendelian mtDNA is transmitted via their direct female line for generations. Elite mares possess superior Mitochondria to transmit to all their female descendants; and thus, these females probably transmit "better" energy levels to their progeny.

The M-factor (effective *mitochondrial* DNA) is something I truly respect, so when advising clients to buy foundation breeding stock, one has to seek fillies and mares from the very best families (from the top 25% of the mare population). There should be many quality stakes winners appearing on the catalog page in the second, third and fourth dams of a potential broodmare, rather than just one big name in black type and very little else. Sometimes the third dam may have produced no black type, due to lack of opportunities with the right matings, so one has to be objective when selecting a good mare. Look for consistency in black type from the female line. Avoid mares from weak families because they inherit and transmit weak conformation.

If we knew which mares possessed the X chromosomes for large heart, plus the M-factor, we could with confidence select foundation mares for a racing and breeding program. The only way one can judge is to guess, using pedigree information at hand, and details of the family's ability to produce stakes winners. This information is available on computers from the Jockey Club's Equiline Services, from Bloodstock Research, or other research oriented organizations.

In closing this chapter, I would like to add brief details about red and white blood cell production.

RED BLOOD CELL PRODUCTION

A hormone called ERYTHROPOIETIN (known in the trade as EPO) regulates erythropoiesis. It is secreted mainly by the kidneys but also manufactured in the liver. This hormone circulates in the blood and upon reaching the bone marrow, stimulates committed stem cells to differentiate. Erythropoietin also shortens the time required for red blood cells to mature. Secretion is inherited by an abnormal increase in the number of circulating red blood cells. It is stimulated by a decrease in oxygen in artificial blood and by deficiency of hemoglobin. Therefore, loss of blood or cardiovascular malfunction that results in reduced amounts of oxygen to the cells causes stepped-up erythropoietin production and *more rapid manufacture of red blood cells.*

Regulation of blood cell production by EPO is thus a homeostatic mechanism for maintaining normal numbers of erythrocytes and an adequate oxygen supply for the cells. The average circulating span of a red blood cell is about 120 days. What worries me most about EPO is that unscrupulous racehorse trainers might use it to boost red blood cells in colts and fillies (thus increasing the amount of red cells for more oxygen to the muscles) and from what I have learned about its effect on young horses, if abused, EPO can be quite dangerous and cause loss of life. Fillies, especially, are affected. Therefore, internationally, EPO should be a banned drug.

WHITE BLOOD CELL PRODUCTION

White blood cells or LEUKOCYTES defend the body against disease causing agents. Unlike red blood cells, white blood cells have nuclei. White blood cells develop from stem cells in the red bone marrow although some complete their maturation elsewhere in the body. The white blood cells enter the circulation by diapedsis. Time spent in the circulation varies from a few hours to many years depending on the type of leukocytes. As white cells wander through the blood system of the body, they phagocytize dead cells, bacteria and other foreign matter. White blood cells are outnumbered 700 to 1 by red blood cells. Granular luekocytes contain potent enzymes that destroy ingested bacteria . Most of the granules are lysosomes.

STORM CAT (Storm Bird-Terlingua by Secretariat), a leading sire of two year olds in North America. He won the Young America Stakes Gr-1 and was defeated a nose to Tasso in the Breeders' Cup Juvenile Gr-1. He has plenty of fast-twitch muscle fibers.

DAHLIA (Vaguely Noble-Charming Alibi) wins the King George VI and Queen Elizabeth Stakes Gr-1 at Ascot for the second time, with Lester Piggott in the saddle. The Queen's classic filly Highclere (J. Mercer) was second, with Dankaro third.

(Above): ROI HERODE, sire of unbeaten The Tetrarch. (Below): LADY JUROR — dam of Fair Trial and Sansonnet, the dam of English champion 2YO colt Tudor Minstrel.

BLANDFORD AND HIS INFLUENCE FOR STAMINA

"The problems of victory are more agreeable than those of defeat, but they are no less difficult."
- Sir Winston Churchill

When a stallion can sire four English Derby winners during his stud career obviously he must possess a degree of prepotency for speed and stamina. In the first half of the century, Blandford (Swynford-Blanche by White Eagle) sired four Derby winners, viz. Trigo, Bahram, Blenheim and Windsor Lad. Another influential classic sire was Hurry On.

In the second half of the 20th century, Nijinsky II (Northern Dancer-Flaming Page by Bull Page) sired three English Derby winners, viz. Golden Fleece, Shahrastani and Lammtarra, as well as being grandsire of another Derby winner, Kahyasi (son of Ile de Bourbon). Nijinsky II also sired French Derby winner Caerleon and North American champion Sky Classic.

Swynford, sire of Blandford, was a product from Lord Derby's Stanley House Stud and was half-brother to successful sire Chaucer, being from English Oaks winner Canterbury Pilgrim by Hermit's staying son Tristan. Blandford's dam Blanche (by White Eagle) was trained at Foxhill where during two seasons in training she failed to show any form. She was unplaced behind Lady Josephine in the Ascot Coventry Stakes and several other juvenile races including the Cheveley Park Stakes. Sent to William Hall Walker's Tully farm in Ireland as a non-winner, Blanche was out of Black Cherry, a half-sister to Ascot Hardwicke Stakes winner Bay Ronald (sire of classic winner Bayardo). Black Cherry came from the Number 3 Family, and traced to Amazon, fourth dam of Pocahontas.

When William Hall Walker (later to become Lord Wavertree) donated forty-three mares and Tully Stud to the government to establish the first National Stud, Blanche was among them. Her first foal was a filly named Bendis, born in 1918 by the stallion Jaegar, which was sold to Spain. Next, Blanche was sent to Newmarket to be covered by St. Leger winner Swynford, and foaled a brown colt named Blandford. Unfortunately this little fellow came down with pneumonia, but came through the ordeal thanks to the kindness of veterinarian Thomas Griffin. Blanche later produced Cesarewitch Handicap winner Seminole (by Diligence), Princess of Wales's Stakes winner Nun's Veil (by Friar Marcus), and the handicapper Silver Hussar (by Silvern).

Blandford was sent to Tully, in Ireland, and as a yearling met with an accident in the paddocks, stripping himself of flesh on both forelegs. His legs healed slowly, but Blandford couldn't make the autumn sales, instead he was sold at the Newmarket December Sales. The brown colt attracted the eye of Samuel Dawson who saw the blemished forelegs, but still liked the yearling by Swynford and bought him for 730 guineas in partnership with his brother, Richard (the renowned trainer, Dick Dawson). It was one of the bargains of the decade. On Wednesday June 8, 1921, Blandford made his debut at Newbury, ridden by P. Mason, in the Kennet 2YO Stakes. The juvenile colt showed early speed and was the first to greet the judge, defeating Lord Derby's Scamp. At Royal Ascot, Scamp confirmed the form by winning the New Stakes by two lengths. At the same meeting Blandford was narrowly defeated in the Windsor Castle Stakes by a colt receiving ten pounds weight, named Alaric. Next morning Blandford was seen suffering from sore shins. He was sent to the paddock to develop. In the meantime, Scamp went on to win the Gimcrack Stakes at York.

Blandford reappeared at three years in late April, winning the Hurst Park Paradise Stakes over one and a half miles, defeating Tracery's son Captain Fracasse, and Spike Island. At his next start, Spike Island won the Irish 2,000 Guineas followed by a splendid triumph in the Irish Derby. Trainer Dick Dawson knew Blandford was not entered for the Derby at Epsom so instead ran the colt at Newmarket in early July. Blandford stormed home to win the Princess of Wales's Stakes over a mile and a half, beating Silver Image by two lengths with Sicyon, well back, finishing third. Silver Image had previously won the Hardwicke Stakes at Ascot, so the form looked smart.

However, Blandford's tendon blew up the next day. Despite careful nursing, his experienced trainer knew that the colt would never race again. Blandford's merits as a racehorse put him among the best three year olds of his career, and had he gone on, may well have been the best of the year. The English Derby was won by Captain Cuttle, by Hurry On. Dick Dawson told everyone about a home gallop at Whatcombe in which Blandford defeated the filly Selene by a length and a half, with Irish Derby winner Ballyheron behind them. The trainer told friends to make use of Blandford which was retired to Dawson's Cloghran Stud in the spring of 1924. That same year, Swynford's son Sansovino won the English Derby; thus, Blandford's pedigree was looking much more appealing.

In his first crop, Blandford sired the more than useful Athford and Buland but it was a colt named Trigo which would set the ball rolling. Bred by William Barnett from a Farasi mare named Athasi, Blandford's son showed trainer Jack Rogers plenty of talent. Trigo ran unplaced before winning the Phoenix Park Plate and Angelsey Stakes but was a disappointment on wet ground in the Railway Stakes. Next year Trigo was sent to the Whatcombe stables and trained for the Epsom classic. The colt won at his debut, scoring in the Berkshire Handicap at Newbury, and then finished ninth in the 2,000 Guineas won by Mr. Jinks. In the English Derby, Trigo was ridden by a twenty year old lad named Joe Marshall, and surprised by winning easily from Walter Gay. Dick Dawson was a happy man at Epsom! The runner-up was a colt by Captain Cuttle.

Trigo won the St. Leger in a desperate struggle with Lord Derby's Bosworth (Son-in-Law out of Serenissima by Minoru). Without doubt, Trigo was the colt which established Blandford's credentials as a sire, and quality mares were sent to him to provide the son of Swynford with more opportunities. Among mares Blandford covered was the Aga Khan's highly strung mare Malva, by the temperamental Charles O'Malley. The result of this union was a colt named Blenheim which would go on to win the English Derby and eventually stand at stud in Kentucky. Blenheim was a successful sire but transmitted temperament. From his European crops he sired Blue Bear (French 1,000 Guineas), Drap d'Or (French 2,000 Guineas), Pampeiro, Wyndham, Mirza II, La Li (dam of Tabriz) and English Derby winner Mahmoud (from Mumtaz Mahal's daughter Mah Mahal).

Federico Tesio sent a mare named Delleana which produced champion Italian three year old Donatello II. At Claiborne Farm, Blenheim became known as Blenheim II, to distinguish himself from another colt of the same name. By the end of 1941, Blenheim was leading sire in the USA. He sired a champion colt named Whirlaway, and a high-class runner like Jet Pilot. Blenheim also sired some good fillies and was to become a successful broodmare sire. His daughter Mumtaz Begum produced Nasrullah (champion English 2YO colt) and Sun Princess, dam of Royal Charger and Tessa Gillian - ancestress of Octagonal, Australia's 1996 champion 3YO colt.

In 1932, Blandford sired thirteen winners and came in seventh on the sires' list. The Aga Khan's Udaipur was the best performer that year, winning the English Oaks. In 1933, Blandford sired numerous winning juveniles including Blazonry (Champagnes Stakes), Campanula (Windsor Castle Stakes) and Mrs. Rustom (Dewhurst Stakes, Gimcrack Stakes). A colt named Windsor Lad was given time to mature for he had an excellent pedigree, being from Resplendent, 1st Irish 1,000 Guineas and Irish Oaks. Windsor Lad had been purchased for 1,300 guineas on behalf of the

Maharaja of Rajpipla by Marcus Marsh, who trained the colt at Lambourn. From only three starts at two, Windsor Lad won the Criterion Stakes at Newmarket. Next year he scored at his reappearance in the Chester Vase and defeated Flamenco a length in the Newmarket Stakes. With Charlie Smirke in the saddle at Epsom, Windsor Lad confirmed his trainer's opinion and won the English Derby by a length over Easton, with a neck to 2,000 Guineas winner Colombo third. Next start, King Salmon and Umidwar were too good for Windsor Lad in the Eclipse Stakes, and an offer of 40,000 pounds was made for Windsor Lad by Martin Benson. The offer was not accepted. It took 50,000 for the deal to be made.

In August at York, for his new owner, Windsor Lad regained his best form to win by three lengths in the Great Yorkshire Stakes and went on to a popular win at Doncaster, taking the St. Leger by two lengths from Tiberius and Lo Zingaro. He equalled Coronach's record time of 3 minutes 1 and 2/5 seconds. At four, Windsor Lad won the Coronation Cup, Ascot Rous Memorial Stakes, and snatched victory in the last strides in the Eclipse Stakes at Sandown, defeating Theft and Fair Trial. Mr. Benson retired Windsor Lad to his Beech House Stud at Cheveley, and the horse only survived until 1943, during the bleakest moments of the war. He sired Windsor Slipper (Irish 2,000 Guineas, Irish Derby and St.Leger), Happy Landing, Phase, and others with talent in the short period of time he was alive.

Blandford's 1934 progeny earnings of 75,707 pounds took him past Stockwell's old record. Badruddin (Blandford-Mumtaz Mahal by Gainsborough-Mumtaz Mahal) was a leading contributor and so too was Umidwar (Blandford-Uganda by Bridaine), both owned by the Aga Khan. Umidwar was to prove a very unlucky performer and his record of 4 wins does not reveal his true racing class. He won the Jockey Club Stakes, Champion Stakes and was runner-up twice in the Eclipse Stakes. Blandford also had a good staying filly out at this time named Dalmary (ancestress of Sadler's Wells and Nureyev). Dalmary, a half sister to that wonderful producer Carpet Slipper, won the Yorkshire Oaks. Kyloe (Blandford-Tabaris) won the Irish 1,000 Guineas and Trigo's little brother Primero dead-heated for the Irish Derby, before taking the Irish St. Leger and later winning the Great Northern St. Leger. 1934 saw the appearance of the Aga Khan's unbeaten champion colt Bahram (Blandford-Friar's Daughter by Friar Marcus) trained by Frank Butters. This good-looking colt with speed was to become a triple crown winner. Winner of 9 races, Bahram won the English 2,000 Guineas, English Derby, St. James's Palace Stakes and the St. Leger. His dam also produced Dastur, runner-up in the English Derby but winner of the Irish Derby and Coronation Cup. Friar's Daughter was out of Garron Lass (Roseland-Concertina by St. Simon) so was a half-sister to the elite mare Plucky Liege. Blandford led the sires list in 1934, and again in 1935 helped by Bahram's exploits.

Blandford's son Bala Hissar, a half-brother to Theft, headed the two year old classifications after winning the Dewhurst Stakes, but this talented juvenile never trained on and was exported to Brazil. In 1937, Blandford finished third on the sires' list behind Solario and Fairway. His Grace (Blandford-Malva by Charles O'Malley) was full brother to Blenheim and much was expected of him. Owned by Mr. James Voase Rank, His Grace was trained by Dick Dawson and captured the Duke of Cambridge Handicap at Newmarket, the Royal Standard Stakes at Manchester and dead-heated at Epsom for the Coronation Cup. His Grace possessed a temperament problem, and had he learned to settle, may have been a good stayer. A very high-class son of Blandford out in 1937 was Midstream, from the Abbot's Trace mare Midsummer. Trained by Ted Leader for Anthony de Rothschild, Midstream won the Newmarket Handicap and Midsummer Stakes, as well as the Stewards' Handicap at Nottingham. He was placed in various important stakes and was purchased by Percy Miller, owner of the famous Kia Ora Stud at Scone in the fertile Hunter Valley of New South Wales, Australia. There, Midstream became the leading sire and got his best Group 1 winners out of Magpie's daughters, thereby reinforcing the influence of Black Duchess.

Blandford's son Pasch, foaled at Banstead Manor Stud, was bred by Mr. H. E. Morris from his Manna mare Pasca, a daughter of Soubriquet, three-parts sister to classic winning filly Fifinella. Soubriquet was by Bayardo's half-brother Lemberg, winner of the English Derby, so Pasch had a pedigree pattern suggestive of a high-class stayer.

Pasch was conditioned by Fred Darling and did not start as a juvenile. The unrefined colt made his debut at Kempton Park, winning the Coventry Stakes, and so impressed were his connections that it was decided to run him in the 2,000 Guineas at Newmarket. Pasch led for the latter part of this mile classic and kept on, scoring a surprise victory over Scottish Union and the Aga Khan's homebred Mirza II. Running as favorite for the English Derby, Pasch could finish only third behind stablemate Bois Roussel (Vatout-Plucky Liege by Spearmint) followed by Scottish Union, but came out next start to take the Eclipse Stakes at Sandown, defeating Fair Copy by three lengths.

Pasch disappointed in the St. Leger, finishing third to Scottish Union and Challenge, but put up a solid performance when second to champion filly Rockfel (dam of Rockefella) in the Champion Stakes. Blandford's son was thereupon retired to Banstead Manor Stud; but he contracted meningitis and died, leaving behind very few foals, the only one of distinction being Pasquinade, dam of the sprinters Royal Palm and Royal Serenade. In 1935, Blandford got Munda, champion filly of Italy.

French breeder Baron Edouard de Rothschild patronized Blandford and was rewarded with La Furka, (dam of Chesterfield which won the Prix Robert Papin and Prix Greffulhe) and Bokbul, winner of the French St. Leger. Reine Isaure (Blandford-Oriane by Sans Souci II) produced Cranach, a stakes winner in Germany which later made a successful sire in France.

The best colt foaled in France at Haras de Meautry was three year old champion Brantome (Blandford-Vitamine by Clarissimus-Viridiflora by Sans Souci II) a close relative of French Oaks winner Rose Vert. In some respects, Brantome might be classed as Blandford's most brilliant stayer. He was certainly as good, if not better, than unbeaten Bahram.

Brantome was a bay colt resembling his broodmare sire Clarissimus, son of a true stayer Radium. Unbeaten at 2 and head of the French 2YO Handicap by an advantage of 8 pounds, Brantome came out at three years scoring an easy win in the important Prix Lupin. Two days before the Prix du Jockey Club (French Derby) he developed a cough and was scratched from the classic.

He resurfaced in the autumn and won the French St. Leger as a prep for Longchamp's valuable Prix de l'Arc de Triomphe. Heavily supported by the stable, Brantome won the Arc by two and a half lengths over Marcel Boussac's French Derby winner Asterus (by Teddy) with English challenger Felicitation a length away third. Here was a real champion and a wonderful sire prospect for French breeding.

In 1935, Brantome easily won the Prix du Cadran by 15 lengths and was soon after sent to England. Unfortunately he was spooked one morning and broke away from his lad, ripping off three racing plates and sustaining bruising to both front feet. Thus, in the Ascot Gold Cup the best he could do was finish fifth to Tiberius. It was a really disappointing effort. Something was wrong.

Taken home to France, Brantome recovered and used the Prix Prince d'Orange as his final test for a second attempt at the Prix de l'Arc de Triomphe. Brantome was expected to win but could only manage fourth to the more than useful filly Samos. The next day his tendon was swollen, and although every effort was made to nurse the horse, he never made it back to the track. Brantome was retired to Haras de Meautry where he became an influential sire, despite having been stolen by

the Germans in 1940. He is the sire of Vieux Manoir who has kept the male line of Blandford very much alive. Exbury is a member of this branch.

Another outstanding French runner sired by Blandford was Mistress Ford (Blandford-Polly Flinders by Teddy) bred by M. Edouard Esmond. This brilliant filly was unbeaten at two in 1935. She won the Prix de Deux Ans at Chantilly, the Prix Robert Papin, Prix Morny, Prix La Rochette, Prix d'Arenberg and defeated talented colts in the Grand Criterium.

At three she was beaten a half length by Blue Bear in the French 1,000 Guineas before winning the Prix de Diane (French Oaks), then scored a desperate victory over Blue Bear in the longer Prix Vermeille. Mistress Ford produced seven winners including Mystery, winner of the Eclipse Stakes with Lester Piggott aboard. She is ancestress of French Derby winner Policeman and Champion Stakes winner Hurry Harriet.

Blandford's son Donatello II maintained the male line in England, siring Alycidon, winner of the Ascot Gold Cup and champion older horse, as well as Crepello, champion 3YO colt of his year, winner of the English 2,000 Guineas, Derby and Dewhurst Stakes. Crepello was a "super" horse.

Blandford died on April 24, 1935 at Whatcombe from the affects of pneumonia. Had he lived, his fee was to be set at 500 guineas. He'd been transferred from Cloghran Stud in Ireland because English breeders wanted access to the stallion. After all, he was the sire of four English Derby winners. In 1938, Blandford led the sires' list for the last time.

Although he was a successful broodmare sire, Blandford never led that list and his best daughter in the long term would prove to be Wild Violet (Blandford-Wood Violet by Ksar). This French mare produced the influential stallion Wild Risk (sire of champion Le Fabuleux) who figures in the pedigrees of Manila, Nashwan and Arazi. Blandford's daughters include Cybiane (dam of English Oaks winner Why Hurry), Simnel (dam of 2,000 Guineas winner Lambert Simnel), Angelus (dam of Nucleus), Morning Dew (dam of four stakes winners, Irish 2,000 Guineas winners Good Morning, Grand Weather, as well as Irish 1,000 Guineas winner Morning Wings and Irish St. Leger winner Morning Madame).

Crepello (Donatello II-Crepuscule by Mieuxce) inherited some weakness to his tendon and might have been a great horse had he not had this problem. He sired Busted, a very good stayer, which in turn sired Bustino, winner of the English St. Leger, Coronation Cup, Classic Trial Stakes, Derby Trial Stakes and York's Great Voltiguer Stakes.

Busted is the sire of Mtoto (dam by Relic's son Mincio) winner of 8 of 17 starts including the Eclipse Stakes Gr-1 twice, defeating Reference Point and Triptych on one of those occasions. He was voted champion older horse of Europe and among other triumphs were the Prince of Wales's Stakes Gr-2 twice, the King George VI & Queen Elizabeth Stakes Gr-1 (defeating Unfuwain and Tony Bin) and the Goodwood Select Stakes Gr-3. Mtoto was gallant in defeat when runner-up to Tony Bin in the Prix de l'Arc de Triomphe Gr-1.

To date, he is the sire of 1996 English Derby winner Shaamit as well as Presenting, Faru and Totostar. The interesting thing about Mtoto is his 3 x 4 linebreeding to Donatello II, grandson of Blandford, and carries more Blandford via Wild Risk (sire of Vimy) and Persian Gulf.

What I find especially puzzling is the fact that the two best male lines of Blandford are through Donatello II and Brantome, both of which carry the influence of Clarissimus! Here now is the pedigree of Mtoto:

				Blenheim
			Donatello II	Delleana
		Crepello		Mieuxce
			Crepuscule	Red Sunset
	Busted			Wild Risk
			Vimy	Mimi
		Sans Le Sou		Court Martial
			Martial Loan	Loan
MTOTO (1983)				War Relic
			Relic	Bridal Colors
		Mincio		Le Pacha
			Merise	Miraflora
	Amazer			*Donatelllo II*
			Alycidon	Aurora
		Alzara		Persian Gulf
			Zabara	Samovar

Persian Gulf (Bahram-Double Life by Bachelor's Double) is of course a well-bred grandson of Blandford, and was himself a good stayer. His daughter Zabara was matched with champion stayer Alycidon, not only a son of champion Donatello II (grandson of Blandford) but linebred to a son and daughter of Blandford's sire Swynford.

Alycidon's daughter Alzara was matched with Mincio from Man O'War's line (thus introducing Rock Sand's influence) and her daughter Amazer was sent to Busted, a stallion carrying not only Donatello II but also a very valuable daughter of Blandford (Wild Violet) to balance Blandford's effect on Mtoto's chromosomes. Here is the pedigree of Alycidon, bred by Lord Derby and winner of 11 races including the Goodwood Cup, Doncaster Cup, Jockey Club Stakes, King George VI Stakes, Ascot Gold Cup. As well, he was 2nd English St. Leger and 3rd Chester Vase. Chaucer is of course out of Canterbury Pilgrim, the dam of Swynford:

				Swynford
			Blandford	Blanche
		Blenheim		Charles O'Malley
			Malva	Wild Arum
	Donatello II			Radium
			Clarissimus	Quintessence
		Delleana		Bridge of Earn
			Duccia di Buoninsegna	Dutch Mary
ALYCIDON (1945)				Bayardo
			Gainsborough	Rosedrop
		Hyperion		Chaucer
			Selene	Serenissima
	Aurora			John O'Gaunt
			Swynford	*Canterbury Pilgrim*
		Rose Red		Marco
			Marchetta	Hettie Sorrel

Aurora, dam of Alycidon, is linebred to Canterbury Pilgrim and her classic winning dam Pilgrimage, so when she was matched with Federico Tesio's homebred Donatello II it duplicated Swynford (4 x 3) and reinforced Canterbury Pilgrim's influence.

It is amazing that three of the most influential male lines in the world today trace to horses bred by Tesio at his Dormello Stud in northern Italy. The three horses he bred and owned, two of which he trained, are NEARCO, DONATELLO II and RIBOT. Two of these colts were unbeaten, and Donatello II was never defeated in Italy.

Nearco sired Nasrullah, Royal Charger, Nearctic and Dante - the four major male branches of the Phalaris tribe. Donatello II sired Crepello and Alycidon - the two major branches of the Blandford tribe. Ribot sired His Majesty, sire of Pleasant Colony, Valiant Nature, Tight Spot and a stallion named Frosty the Snowman which could get many good runners if breeders would only appreciate his pedigree.

Pleasant Colony (His Majesty-Sun Colony by Sunrise Flight) is linebred to Mahmoud, and was champion three year old of North America, winning 6 of 14 starts for $965,383 including the Kentucky Derby Gr-1, Preakness Stakes Gr-1, Wood Memorial Stakes Gr-1, and Woodward Stakes Gr-1. He is the sire of St. Jovite (champion, 1st Irish Derby Gr-1, King George VI & Queen Elizabeth Stakes Gr-1, Angelsey Stakes Gr-3), Pleasant Tap (champion, 1st Jockey Club Gold Cup Gr-1, Suburban Handicap Gr-1, Malibu Stakes Gr-2 and 2nd Breeders' Cup Sprint Gr-1), Pleasant

BLENHEIM (Blandford-Malva) returns to scale after winning the English Derby for his owner-breeder, the Aga Khan. He sired Mahmoud, Donatello II and Nasrullah's dam Mumtaz Begum.

Stage (champion filly), Colonial Affair (1st Belmont Stakes Gr-1, Jockey Club Gold Cup Gr-1), and Roanoke (Californian Stakes Gr-1). Pleasant Colony has sired his best runners from mares that carry Nearco and Hyperion, two ancestors linebred to St. Simon.

Blenheim, winner of the Derby, stood in France at the Aga Khan's Haras Mary-la-Ville until the completion of the 1936 covering season, and was then sold to Kentucky for 45,000 pounds. His best European runner was Mahmoud (Blenheim-Mah Mahal by Gainsborough) winner of the Derby in record time. Mahmoud became leading sire in North America and his sons include the successful sires The Axe II and Cohoes. Mahmoud is a rather good example of a thoroughbred inheriting *Arabian phenotype* and sired two of the most influential mares in North America's stud book, viz. Almahmoud (grand-dam of Northern Dancer, Arctic Tern and Halo) and Grey Flight, a champion filly which produced stallion sons. Almahmoud is the inbred mare found in the pedigree of Coup de Folie, dam of French champions Machiavellian and Coup de Genie, and of Group 1 winner Exit to Nowhere.

Blandford will remain a genetic force for stamina. He was responsible for correcting the conformation of many well-bred mares that were flat in the knees. Blandford was "over" at the knees, and so were many of his progeny.

COUP DE FOLIE (by Halo) photographed in Kentucky. This Group winner inbred to Almahmoud, produced two champions and is one of the many mares owned by the Niarchos family.

In Australia, Stardust's son Star Kingdom was not especially perfect in his forelegs nor sound in his sesamoids but when mated with mares from Blandford's line sired Group 1 winners. I suppose one of the most valuable aspects about Blandford's pedigree is his connection with the high-class broodmare Black Cherry because she also appears in the pedigree of Teddy and Bayardo (sire of Gainsborough). Almost every time these strains were mixed it produced an upgrading for classic speed, so one has to wonder about any special impact from the mare Black Duchess.

I design matings for clients which duplicate Blandford and his father Swynford because I know this line can supply soundness in the knees. The only weak trait handed down from Swynford's sire John O'Gaunt is tendon weakness, something which hindered Crepello after he won the English Derby, and bothered Brantome after he won the Prix de l'Arc de Triomphe. Blenheim, his son, inherited temperament from Charles O'Malley, not from his sire. Descendants of Blandford in the main possess courage, heart, and are genuine on the racetrack. For these reasons the name of Blandford will forever be remembered, thus it was nice to see his descendant Shaamit win the 1996 English Derby, thereby continuing the history of this wonderful classic line.

PURCHASING FOUNDATION BREEDING MATERIAL

- BUYING FROM PUBLIC AUCTIONS

" A good archer is not known by his arrows, but by his aim."
- Ralph Emerson

If one is inexperienced with the genetic/monetary value of breeding or racing stock, my humble advice is to seek a professional, be it a reputable bloodstock agent (reputable agents are honest since they are in the business for the long haul), an experienced trainer, a reliable farm manager or a thoroughbred breeding consultant. I firmly believe there are many stud farm managers which have enormous experience to offer any newcomer into the business. Certain guys I've come to know well are hands-on managers like Tim Richardson (Haras Fresnay-Le-Buffard, France) and his brother Chris Richardson (Cheveley Park Stud, Newmarket), Peter Orton (Arrowfield Stud, Australia), Duncan Taylor (Taylor Made Farms), Irish-born Clifford Barry (Pin Oak Stud), Jim Cannon (Overbrook Farm), Irish-born Michael Hernan (Gainesway Farm), Bill Landers (Hermitage Farm), Irish-born Peter Kirwin (Spendthrift Farm), John Vincent (Collingrove Stud, Australia), Bobby Spalding (Elmendorf Farm), Irish-born Garrett O'Rourke (Juddmonte Farm), Pope McLean Jr. (Crestwood Farm) and Derek Field (Widden Stud, Australia). Other professionals I respect who have the knowledge to give to owner-breeders are people like Ted Carr (Brookside Farm), Gus Koch (Claiborne Farm), Dan Rosenberg (Three Chimneys Farm), Rich Decker (Prestonwood Farm) etc. The list becomes too long to name every "quality" farm manager around the globe. Next we have talented veterinarians managing farms, e.g. Dr. Barry Simon at Ashford Stud and South-African born Dr. John Chandler, general manager of Juddmonte Farms in Kentucky. You might wish to go to the "top" and speak with Alice Chandler (Mill Ridge Farm), Josephine Abercrombie (Pin Oak Stud), Robert Clay (Three Chimneys Farm), Will Farish (Lane's End Farm), Johnny Jones (Walmac), Brereton Jones (Airdrie Stud) Arthur Hancock and his brother Seth Hancock. These people have a wealth of knowledge and really are approachable.

Joss Collins (BBA) is probably the best judge of a yearling I've ever come across and although I don't know him personally James Delahooke is a guy I respect, for he has made some brilliant yearling purchases in the past decade. Mike Ryan is another. I've come across a lot of good judges of broodmares and as luck would have it, I was taught by the best - Sir Philip Payne-Gallwey, a director of the British Bloodstock Agency and consultant to the National Stud in England. It was Philip who first introduced me to "port" and I have enjoyed many a glass or two, ever since. Sir Philip was for many years racing manager for Stavros Niarchos, owner of a strong racing stable in Europe. He handed over that role to a much younger man, Alan Cooper. Not many farms employ pedigree consultants, but this is something that probably will eventuate on a grander scale next century. To run a professional business, a commercial farm should offer a wide spectrum of service to clients. Purchasing breeding stock is one of the toughest assignments for a thoroughbred breeder. Constraints of location and investment bankroll are but two tiny factors governing the way one approaches an introduction to the racehorse breeding business.

Should I own a farm or board my mares? Should I go for quality stock with less numbers to feed or should I play the numbers game ? What are my long-term objectives? Do I want to become a commercial breeder, a private breeder, or might I perhaps develop a parallel program and try to do

both successfully? Many times I've advised clients they could do both by setting-up two separate programs, each with different disciplines. Unless you specifically want to own a farm with its innate overheads, try boarding your breeding stock for those first few primitive years because you will feel comfortable in the knowledge that your horses are well managed, and time will enable you to gain experience and learn the ins and outs of the breeding industry.

I will outline a basic philosophy for the breeder wishing to either breed for the market or for the racetrack (the person who breeds to race). Even though their short-term goals may differ, both the commercial and private breeder should treat their investment as a business, rather than a hobby.

The prospective commecial breeder needs to establish a working budget, hire an experienced advisor or agent, plus a veterinarian, and thereupon set about buying fillies or mares from strong black type families, especially those with representatives of Group/Graded stakes class on the catalog page. Cheap black type on a catalog page would be a mistake! Learn the difference as quickly as you can between *quality* black type and *cheap black type.* The best quality breeding stock are always entered at major sales like Tattersalls at Newmarket (England), William Inglis and Son's Easter Sale (Sydney, Australia), Wrightson's Karaka Breeding Stock Sale (Auckland, New Zealand), and Keeneland's and Fasig-Tipton's November Sales (Lexington, Kentucky).

Your advisor/agent can recommend quality breeding stock or yearling fillies. Your veterinarian can thoroughly examine them prior to sale-time. Yearlings should be scoped to ensure clean air-flow exists because *caveat emptor* exists at horse sales - you buy at your own risk. Perhaps you may not be aware of the increasing number of yearlings turned back by buyers as a result of failing to scope cleanly. Let's face it, nobody wants to begin a program with stock possessing genetic problems, no matter how well-bred they might be. When selecting and buying yearlings, choose those by proven sires, or else if these are too expensive, take a risk on first or second season sires that are sons of powerful mares. Don't get carried away by a stallion's sire line: concentrate more on the *dam of the stallion* which sired the yearlings you like. It is he who carries the important X factor to deliver to all his filly foals. Examine the family's history - are there examples of quick maturing performers?

When purchasing broodmares, one needn't seek the perfect specimen in terms of conformation, although this would be nice. Look for *health* first, *quality* second, and *soundness* third. Select mares from respectable stakes producing female lines, preferably sired by sound stallions that can make successful broodmare sires - better still, by *proven* broodmare sires. The kind of mare to look for should have fairly correct legs, good feet, good shoulder angle, sufficient bone, enough size and plenty of space in the pelvic area. Your veterinarian should check the health records and history of any mares on your short-list, examine them for breeding soundness pre-sale time, and explain to you any apparent physical strengths or weaknesses.

He or she must check for pregnancy to make certain the mare you wish to purchase isn't carrying twins. The shape of the mare's vulva will indicate whether she will be the type to get "dirty" or not. Mares with sunken vulvas should be avoided unless you are a true risk-taker. However, there is one conformational aspect I completely ignore when it comes to judging a broodmare. I am one of those who will not penalize mares that are pidgeon-toed, providing they are within the realms of acceptance. Many outstanding mares have this fault, sell cheaper than they should, yet produce very correct yearlings. "Toe-in" is usually the result of a recessive gene being expressed and maybe only one foal out of four will likely inherit it. A mare's breeding record will show up any years when she went barren, absorbed or aborted a foal, or produced a dead foal. Sometimes her earlier progeny may have met with an accident, causing the foal to be a non-runner. You need details to make allowances for something like this. You must do your research well. Also, please run a computerized produce record prior to sale day so as to check on any updates of foals that might have won since the catalog was printed. Vendors usually supply updates, but to be on the safe side, do your own research (in case the vendor is unaware of a very recent stakes winner from the family).

I suggest you seek the opinions of professional people because they will have their own opinions, and what's more, good reasons for forming them. As John Milton once wrote: "The opinion of good men is but knowledge in the making". If we all shared the same opinion, we'd all marry the same lady, or the same guy, whichever is applicable. What I write may seem basic to many readers but for the newcomer into the industry, there may be one or two points that might help. To be truthful, there can be no "experts" in the horse-breeding business. We are all amateurs, trying to get ahead. Just when you feel confidence is at a high, something will happen and shoot you down in flames. At least, that's the way I feel about the subject. One has to have a stout heart and learn to be a good loser as well as a good winner. Everyone is humbled at some time or another in this business.

At a recent seminar for TOBA members (the Thoroughbred Owners and Breeders Association of North America) I was asked to be on a pedigree panel with Sissy Woolums (Sissy is what I call a real expert since her particular forte is expertise re international cataloging standards), Bill Landers (that likeable guy from Hermitage Farm) plus breeding journalist, racehorse owner-breeder, and novelist, Ed Anthony. We were asked to estimate the value (sight unseen) of a few yearlings entered for the Keeneland sales, and would you believe it, our appraised values for each colt or filly varied considerably. We shared differences of opinion. That's fine. It's healthy, but it shows how difficult it is for a professional person to *appraise a yearling's value.* It is much easier to appraise a broodmare's estimated value. Some commercial farms even go so far as to get two separate appraisals to help estimate the value of yearlings and breeding stock, e.g. for insurance purposes.

In the USA, I advocate horse-breeders join TOBA. This can be done by contacting TOBA's representative: Helen C. Alexander, Thoroughbred Owners and Breeders Association, P.O. Box 4367, Lexington, Kentucky, 40544 or phone for information on (606) 276 2291. This organization is there to help thoroughbred owners and breeders. TOBA seeks to educate and help experienced and inexperienced members, and I highly recommend it. Sissy Woolums by the way, is the president of Pedigree Associates, a company that provides an invaluable service for cataloging standards.

Prior to a sale, one must research the sales catalog. Printed on the catalog page is information about sire, dam, grand-dam, and sometimes great grand-dam. All produce of a mare may not show up on the page, so one has to do a bit of homework. Commercial sellers edit their catalog page with great care, highlighting the best, and shunting superfluous information out of the way. This is of course ethical. Their aim is to have all the most important information on the catalog page and if it means editing a few lines to add more black type at the bottom of the page (say for the great grand-dam) then so be it. It is the vendor's prerogative to edit information: therefore, the buyer must find out what's missing on that page.

Soundness is something I look for when reading the produce record of dams on a catalog page. By this, I mean I like to see winners of more than 6 races rather than plenty of winners of just one race. I prefer Group or Graded stakes runners to Listed or non-listed stakes because this shows the horse's relatives have performed against the very best company. I look for any runners which set track records. I analyze the quality of sires used to get the mare's previous winners. Were they by cheap or expensive stallions? Mares or yearling fillies from poor fertility lines will have female ancestors that didn't produce many foals. Be alert to a spotty produce record. Families with just one big stakes winner on the page indicates the female line is not consistent in its ability to produce stakes runners. Mentally I can extend most three-generation pedigrees printed on the catalog page, noting any significant duplication of superior ancestors (this comes from years of practice, but my memory isn't as good as it used to be). Newcomers may need to run a 5-cross pedigree extension and study it. Once a filly or mare is purchased, please know about your rights as per the payment of accounts, letters of credit, and especially *Conditions of Sale* which are printed at the beginning of catalogs. Read these conditions with great care. If a mare is sold as being pregnant and you check and find she is not pregnant on the same day as the sale, you probably have the option to hand her back or negotiate with the vendor, but be aware of the *Conditions of Sale.* That's why TOBA is such a great organization. They teach people the do's and don'ts. Finally, one must purchase quality breeding stock that fits your program - speed or stamina, or both. Decide early, so that your program can benefit. After all this, then decide about insurance. Will you self insure and take the risk? Or will you cover your investment? That's a personal decision you must make.

There are excellent sources to help match your mares and fillies, like The Blood-Horse's book about breeding "nicks" (what lines suit each other and supply stakes winners) or you may seek the advice of bloodstock agents, trainers or pedigree consultants. I find it invaluable to chat with stud farm owners and their hands-on managers. They are there to help us, so don't be shy. The British Bloodstock Agency has an excellent pedigree department for Europeans and service their clients with mating advice. Purchase thoroughbred books and read about the names in pedigrees which fill the ancestry of your fillies and mares e.g. "The Great Stallion Book" by Richard Ulbrich. Keep learning, for knowledge is power. Remember, we die still learning about racehorse breeding.

Certain people in the industry will say that it doesn't matter once you look beyond a three-generation pedigree. Let me state categorically, those people are downright ignorant! I don't tolerate lazy people easily, but I will excuse uneducated people. A pedigree is a truthful blueprint of the animal's ancestry, revealing heaps of information about racing class, fertility, soundness, temperament, genuineness, courage, speed and stamina. The first three generations show the parents, the grand-parents, and the great-grandparents (numbering 12 immediate ancestors), but where do they come from? Do they not have ancestors themselves? They certainly don't come from thin air! Is it worthwhile to analyze the fourth generation (the fourth remove) and the fifth generation (the fifth remove) of a pedigree to gain more information about a horse's ancestry? The answer is very definitely, YES.

Racehorses invariably have ancestors duplicated. I *prefer* yearlings or mares that have *sons and daughters* of the most influential ancestors *duplicated* in their pedigrees. I need to know what is behind the immediate ancestors. So should you. Some pedigree patterns don't turn me on because they are bred the wrong way for a runner. When I see yearlings with pedigrees revealing a nice balance of sons and daughters of a great sire, or a great mare, or both, I get excited. This technique of using both sexes of an ancestor helps balance *genetic influence* in such a way as to hype up speed. It deploys influence on both sides of chromosomes. I've been studying international pedigrees for more than forty years and consider myself a specialist. During the past eight years I've helped clients design the matings of champions and almost 140 individual stakes winners from a limited number of matings. John Prather and Monty Hinton, friends who excel at pedigree research, and guys like Alan Porter and Les Brinsfield are specialists like me who explore new genetic affinities between lines. (Three Chimneys Farm has on staff a very experienced full-time pedigree specialist too). We do this via painful hours of pedigree analysis to try to be one step ahead of our competitors. I'm sure there are others who do the same. Alec Wildenstein, for example, is quite clever at designing special patterns. He and his father create matings purely for the racetrack, sometimes using unpopular stallions, and they get results. However, if one is breeding commercial yearlings, you must give buyers what they want and thus fashion comes into the formula. Pedigrees are like gene maps - they reveal good and bad, and can indicate "hidden" traits.

With regard to buying shares and seasons in stallions, this is another personal decision that must be made. Seek advice from professionals and you will likely receive different opinions. If you are able to guess which young stallions will become influential, then by all means buy shares, because in the long run, they are nearly always more economical than purchasing seasons (nominations) every year. From owning a share in a stallion, it doesn't mean you have to use the season. It can always be traded for one in another stallion.

Perhaps the best investment of all is to buy shares in *proven sires,* even if it means paying slight premiums. The next best investment has risk, but the upside can be overwhelming, and that is to buy a share in a young stallion that was a champion two year old, or in a Group/Graded winning son of a great racemare or producer.

One has to have a crystal ball in this business. The stallion market is like the futures market. Guess right and you could be on "cloud nine": guess wrong, and it'll cost you money. If your guess fails, accept your fate and begin again. From my experience, some owners are much luckier than others. Why is this so?

HEIGHT OF FASHION, co-champion English 2YO Filly. She is the dam of Nashwan, Europe's Horse of the Year, winner of the English Derby and 2,000 Guineas and of Unfuwain and Mukddaam.

SUCCESSFUL EXAMPLES OF INBREEDING

" If a man will begin with certainties, he shall end in doubts; but if he will be content to begin with doubt, he shall end in certainties."
- Francis Bacon.

The development of the modern racehorse was fairly rapid in the 18th century because the British are renowned for their stubborn, perhaps tenacious ability to improve domestic animals. They are intent on being successful at animal breeding. They improved many breeds of cattle, sheep, birds, so why not create a hybrid horse and call it a "Thorough-bred"?

I am not suggesting the British are the only ones who can do it, but I do believe in giving credit where credit belongs. Mind you, the British had material worth using to establish the breed known as the racehorse. Quality stallions and mares were imported from Persia, Egypt and Turkey which were the results of breeding policies undertaken long ago by bedouin tribes and the royal studs of Persian kings. The *Arabian* was a purebred, a horse of endurance, quality and refinement which had taken many hundreds of years to be purified via *selective breeding* by middle eastern horse-lovers to reach such perfection.

British horse-breeders decided they wanted more size, more speed, and even more refinement in their horses, so they introduced other strains to mingle with the genes of the purebred *Arabian.* Among pioneers of the modern racehorse was James D'Arcy, manager of the royal stud farms for King Charles II. This gentleman must have been a skilled animal breeder because he bred so many worthwhile stallions and mares that luck could not have played the dominant role. Nobody could be that lucky! Another important gentleman was Tregonwell Frampton who made a lasting contribution.

From a relatively small gene pool, British breeders enhanced and shaped the breed into what it is today. Along the way, it was necessary to *inbreed* or *linebreed* to the best individuals which transmitted the traits or characteristics that were desired in a racehorse. This too, was *selective breeding.* Horse breeders attempted to "fix" certain favorable genes in their breeding stock. One must understand that transport was a problem in the 18th and 19th centuries, where roads could be washed away, and having to walk stallions from place to place was common. This meant that the most successful sires were patronized to the fullest, because of restrictions. Inbreeding to *The Darley Arabian, The Byerley Turk, The Godolphin Arabian* and to their famous male descendants Matchem, Herod, and Eclipse provided instant upgrading for speed, albeit over distances of two miles or more. Gradually inferior families were unable to compete with the better lines, and faded into obscurity.

Herewith are examples of class racehorses possessing pedigrees either designed by their breeders by accident, or on purpose, that were *inbred within three generations: (a)* the intensely inbred Barcaldine (Solon-Ballyroe), Flying Fox (Orme-Vampire), Wellingtonia (Chatanooga-Araucaria), Paulowitz (Sir Paul-Evelina), and *(b)* the "almost" intensely inbred Sainfoin (Springfield-Sanda), Amphion (Rosebery-Suicide), Boston (Timoleon-mare by Ball's Florizel), Cap-a- Pie (The Colonel-Sister to Cactus), Gohanna (Mercury-mare by Herod), Galopin (Vedette-Flying Duchess), Orville (Beningborough-Evelina), Saltram (Eclipse-Virago), La Saggittaire (Le Sancy-La Dauphine), Priam

(Emilius-Cressida), Sorceror (Trumpator-Young Giantess) and Wisdom (Blinkhoolie-Aline). These names may not mean much to a young reader, but they were pillars of the best pedigrees of the past.

Triple crown winner Flying Fox (1896) by Orme out of Vampire by Galopin, was incestuously inbred 3 x 2 to Galopin via a son and daughter. Galopin's sire Vedette was inbred 3 x 4 to Blacklock (12 wins from 15 starts) via a son and daughter. Galopin himself was inbred 3 x 3 to the high-class horse Voltaire (5 wins from 6 starts including the Doncaster Cup) and linebred 5 x 5 to Phantom. Here is Flying Fox's pedigree:

		Ormonde
	Orme	Angelica by *Galopin*
FLYING FOX		*Galopin*
	Vampire	Irony

The third foal of his temperamental dam, Flying Fox won 9 races from 11 starts including the English 2,000 Guineas, Derby (defeating Domocles), St. Leger, Eclipse Stakes, Jockey Club Stakes, and Princess of Wales's Stakes. He became a leading sire in France, where he sired unbeaten Ajax, the sire of French champion three year old Teddy, sire of Sir Gallahad III, Bull Dog, Asterus and Sun Teddy. Flying Fox was bred by the Duke of Westminster who did not wish to travel his vicious mare Vampire (she had mauled a stablehand) so sent her to his own stallion Orme.

Galopin (1872) won 10 of 11 starts and was a high-class racehorse whose wins include the English Derby, Newmarket Derby, Fern Hill Stakes. His inbreeding to Voltaire might have been significant, but his pedigree also revealed *linebreeding* to great sire Orville and his influential full sister Orvillina. Galopin was leading sire in England and was a sire held in the very highest regard. Apart from St. Simon, unbeaten and one of the most prepotent sires in the history of the thoroughbred, Galopin sired Galliard, Disraeli (both 2,000 Guineas winners), Donovan (English Derby and St. Leger winner) and Galicia, dam of Bayardo and English Derby winner Lemberg.

On the opposite page is the pedigree of St. Simon, showing the closeness of three-quarter sister and brother Merope and Velocipede (3 x 5), meaning Merope in the third generation of St. Simon's pedigree, and Velocipede in the fifth generation. Whereas Merope is by a son of Blacklock, Velocipede is by Blacklock. Furthermore, the duplication of Orville and his sister Orvillina is supported by their half-brother Paulowitz, thus reinforcing their dam Evelina (maternal strength).

St. Simon zig-zags in *alternate sex line* from his dam St. Angela to King Tom to Pocahontas, and may have inherited his X chromosome from Pocahontas. His Y chromosome definitely came from his sire Galopin. It may be stretching the imagination a little, but perhaps St. Simon inherited a large heart from Pocahontas's input.

Certainly the genetic influence from Blacklock (Whitelock-mare by Coriander) must have been emphatic. This half-brother to St. Leger winner Theodore was a colt foaled in 1814 and raced four seasons for 12 wins from 15 starts. He was a good two year old winning both his starts at that age, and was runner-up to Ebor in the English St. Leger, beaten a short neck. He won the Doncaster Cup, Doncaster Stakes, and York's Great Subscription Sweepstakes. Standing 15.3 hands high, he was one of the biggest stallions at stud in the early 1820's and transmitted sickle shaped hocks, a characteristic which St. Simon inherited and passed on to his descendants.

The next interesting example of inbreeding is Ayrshire (Hampton-Atalanta by Galopin), winner of 11 of his 16 starts and sire of Robert Le Diable and Gas, dam of the brilliant Cicero. Ayrshire won five races at 2 years including the Champagne Stakes and Chesterfield Stakes, won the English 2,000 Guineas, Derby and Riddlesworth Stakes at three, and the Eclipse Stakes at four.

					Blacklock
				Voltaire	mare by *Phantom*
			Voltiguer		Mulatto (dam by *Orville)*
				Martha Lynn	Leda
		Vedette			Sir Hercules
				Birdcatcher	Guiccioli
			Mrs Ridgway		Inheritor
				Nan Darrel	Nan by *Blacklock*
	Galopin				*Sultan* by *Selim*
				Bay Middleton	Cobweb * by *Phantom*
			The Flying Dutchman		Sandbeck (*ex Orvillina)*
				Barbelle	Darioletta (dam by *Selim)*
		Flying Duchess			*Blacklock*
				Voltaire	mare by *Phantom*
			MEROPE		*Juniper*
				Merope's dam	*mare*
ST. SIMON					Whisker
(1881)				Economist	Floranthe
			Harkaway		Nabocklish
				mare	Miss Tooley
		King Tom			*Sultan* by *Selim*
				Glencoe	Trampoline *
			Pocahontas		Muley by *Orville*
				Marpessa	Clare
	St.Angela				Paulowitz (ex *Evelina)*
				Cain	mare
			Ion		Edmond by *Orville*
				Margaret	Medora by *Selim*
		Adeline			VELOCIPEDE by *Blacklock*
				Hornsea	mare by Cerberus
			Little Fairy		Zodiac
				Lacerta	Jerboa

PEDIGREE NOTES:

* Web is the dam of Trampoline (dam of Glencoe)
* Web is also dam of Filagree, dam of Oaks winner Cobweb

MEROPE is a 3/4 sister to VELOCIPEDE as follows:

Merope by Voltaire (son of BLACKLOCK) from a mare by Juniper
Velocipede is by BLACKLOCK from mare by Juniper (Merope's dam)

Orville and Orvillina are full brother and sister

Evelina is the dam of Paulowitz (sire of Cain), Orville and Orvillina.

Ayrshire has a rather unique pedigree pattern because he carries four sets of brothers and sisters in his make-up. Here is his pedigree:

				Touchstone
			Newminster	*Beeswing*
		Lord Clifden		Melbourne
			The Slave	*Volley by Voltaire-Martha Lynn*
	Hampton			*Rataplan (brother to Stockwell)*
			Kettledrum	Hybla
		Lady Langden		Lanercost (by son of *Tramp)*
			Haricot	Queen Mary
AYRSHIRE				*Voltiguer (g'dam by Orville)*
			Vedette	Mrs Ridgway
		Galopin		Flying Dutchman (to *Orvillina)*
			Flying Duchess	Merope by *Voltaire*
	Atalanta			Windhound
			Thormanby	Alice Hawthorne
		Feronia		*Stockwell*
			Woodbine	*Honeysuckle (Touchstone-Beeswing)*

PEDIGREE NOTES:

1) Newminster is full brother to Honeysuckle
2) Rataplan is full brother to Stockwell
3) Volley is full sister to Voltiguer
4) Orville is full brother to Orvillina - dam of Sandbeck in The Flying Dutchman's pedigree.
5) Lanercost is by Liverpool, son of Tramp whereas Stockwell and Rataplan carry Tramp via Glencoe, sire of their dam Pocahontas.
6) Stockwell and Rataplan carry 2 full brothers and their full sister.
7) Stockwell's sire The Baron traces to Gadabout by Orville.

Imagine the concentration of these high-class ancestors and the effect they might have to increase racing class. Ayrshire's pedigree has a lot of genetic strength.

Multiplication of the finest qualities of the best performed or best bred thoroughbreds of each generation is the ideal objective to "fix" desirable characteristics. Dilution of quality strains using a technique called *outcrossing* (i.e. non-duplication) may have success initially through hybrid vigour, but will weaken prepotency for subsequent generations. Other examples of collecting genetic strength via the method of *inbreeding* or close linebreeding was evident in the pedigrees of Stockwell and Rataplan (4 x 4 x 4 to full siblings Whalebone, Whisker, and their sister Web); the famous mare Diversion (inbred 2 x 3 to Whalebone and his sister Web), she being the grand-dam of Rosicrucian and The Palmer (full brothers found in the pedigree of Hyperion); the famous mare Selene (inbred 3 x 4 to classic filly Pilgrimage), she being dam of Hyperion, Sickle, Pharamond II; The Tetrarch, unbeaten English 2YO champion, inbred 4 x 4 to Rouge Rose and 4 x 4 to Doncaster; Vahren, dam of The Tetrarch, she being inbred 4 x 4 to Pocahontas's son King Tom; Americus, inbred 2 x 2 to Norfolk and his full sister The Nun - Americus sired Lady Americus, dam of famous mare Lady Josephine (ancestress of Nasrullah, Royal Charger, Mahmoud); Irish Lass, winner of the

Prix de Minerve Gr-3 and inbred 3 x 3 to Nogara via her sons Nearco and Niccolo dell'Arca - Irish Lass is dam of Irish Ball (Irish Derby) and Irish Bird, dam of Bikala (French Derby); Crafty Admiral, inbred 2 x 3 to Sir Gallahad III and his full brother Bull Dog - he appears in the pedigrees of Danzig and Affirmed; Gallant Man, inbred 3 x 3 to Mah Mahal and 4 x 4 to Blandford - winner of 14 races including the Belmont Stakes, Hollywood Gold Cup and sire of 44 stakes winners; Tudor Minstrel, champion English 2YO colt and winner of the 2,000 Guineas with inbreeding to Gondolette; Reigh Count, 11 wins including Kentucky Derby, inbred 3 x 3 to St. Frusquin, and sire of champion Count Fleet which is linebred 4 x 5 to Sierra and her brother Sainfoin; Heliopolis, inbred 4 x 3 to Canterbury Pilgrim, 5 wins including Princess of Wales's Stakes, Chester Vase, and a leading sire in USA; Spy Song, inbred 3 x 3 to Peter Pan and linebred to Sierra and Sainfoin, 15 wins 1st Arlington Futurity, Chicago Handicap, 2nd to Assault in Kentucky Derby; High Time, inbred 3 x 2 to Domino, a leading sire in USA; and Havresac II, inbred 2 x 3 to St. Simon with St. Simon's sister further back in the pedigree - leading sire in Italy, and broodmare sire of Nearco.

A pedigree which fascinated me was that of Sarazen, 27 wins and earner of $225,000 when stakes were low. He won 10 consecutive races at 2 years in the USA including 6 stakes races. At 3, he won 8 of 12 starts including the Third International Stakes (defeating Epinard) going 1 1/4 miles in 2-00 4/5 which set a new American record time for the distance. He had some temperament, but retired sound. Here is his pedigree:

			Commando by *Domino*
		Ultimus	Running Stream by *Domino*
	High Time		*Domino*
		Noonday	Sundown
SARAZEN			Order by *Bend Or*
		Box	*Angelica (sister to St. Simon)*
	Rush Box		Singleton (*St. Simon-mare by Bend Or)*
		mare	mare

Sarazen's pedigree shows he is the product of two inbred parents, High Time (inbred three times to champion Domino, via two sons plus a daughter) and Rush Box, (inbred to Angelica and her full brother St. Simon, and to the great Bend Or). Here is an instance of hybrid vigour coming via two different inbred strains. There must have been a positive effect because Sarazen was an outstanding racehorse, especially as a juvenile. Bend Or is an ancestor which could have transmitted left laryngeal paralyis, a recessive factor that can skip generations before it is exposed. Some of Domino's progeny bled. Sarazen did not inherit either weakness. Let's now look at the pedigree of the chestnut mare Sanda (1878), dam of English Derby winner Sainfoin and of Sierra, dam of champion sprinter Sundridge. Sanda established a genetically powerful damline and is ancestress of the world's richest earning racehorse, Cigar.

			Newminster
		Lord Clifden	The Slave
	Wenlock		*Rataplan*
		Mineral	Manganese
Sanda (1878)			*The Baron*
		Stockwell	*Pocahontas*
	Sandal		Don Juan
		Lady Evelyn	Industry

Here we see inbreeding to full brothers Stockwell and Rataplan, sons of Pocahontas.

Sainfoin (Springfield-Sanda) bred by Queen Victoria, was a classic winner sold at the annual yearling sale from Hampton Court. He was a touch over 15.2 hands high and won the 1890 English Derby from Le Nord and Orwell. A symmetrical colt, he caught the eye of trainer John Porter and Sir Robert Jardine when they attended the royal stud's sale and purchased Sainfoin for 500 guineas, a considerable sum in those days, so he must have been a fairly correct youngster. One of the reasons why they chose to buy the yearling was the fact he was inbred 3 x 3 to the great Stockwell, and had a line of his full brother Rataplan in the pedigree.

A rich chestnut in color, Sainfoin won his only start at 2 years in the Astley Stakes at Lewes in August, 1889. He did not race again until the Sandown Esher Stakes, carrying a light weight, which he won by 4 lengths from 12 starters. This victory so impressed a rich young man named James Miller, that he offered 6,000 pounds to Sir Robert Jardine, which was accepted. For his new owner, Sainfoin won the Chester Dee Stakes, scoring a half length from Bull's Eye, and was heavily supported to win the Derby, starting second favorite at 100/15. It rained at Epsom, and Sainfoin relished the going. He won the Derby by 3/4 length from odds-on favorite Surefoot which had earlier won the 2,000 Guineas. Next time out Sainfoin was defeated in the Ascot Hardwicke Stakes by Amphion, with Surefoot third. After a let-up he was set for the St. Leger, but could only manage fourth to that crack filly Memoir. It was rumored he had a suspensory ligament problem. At his final start at 3, he was unplaced in the Free Handicap, pulling up lame. He was lightly raced at 4 and 5, being stakes placed under heavy weights, but was never right after his Derby win. He sired triple crown winner Rock Sand, the sire of champion Tracery.

Sainfoin's sire Springfield was a very handsome bay colt by Stockwell's chestnut son St. Albans. The dam of Springfield was Viridis, by Touchstone's grandson Marsyas. Springfield was bred by Queen Victoria in 1873. Bought by Mr. Houldsworth as a yearling for 320 guineas, the colt fractured his pelvis, but made a swift recovery. In workouts he showed lots of speed. At 2, Springfield won the Prince of Wales Plate at York and the following day, won the Gimcrack Stakes by 3 lengths. Later he was second in the Criterion Stakes and again runner-up when second to Kisber in the Dewhurst Stakes. (Kisber came out and won the English Derby the following summer).

At 3, Springfield was unbeaten in 9 starts, his triumphs including the Fern Hill Stakes, Ascot Biennial Handicap, Newmarket July Cup, Stockbridge Handicap and the Free Handicap. At 4, he confirmed he was England's best sprinter, winning the July Cup again, the Champion Stakes, All-Aged Stakes, and Queen's Stand Plate. By the end of his career he had notched up 18 of his 20 starts for 6,720 pounds earnings. He retired to stud at 100 guineas which rose to 200 guineas upon his success as a sire. He died aged 25 years on March 31, 1898. Apart from Derby winner Sainfoin, he is best known as the sire of Briar-root (English 1,000 Guineas) and Watercress (Prince of Wales's Stakes, Hardwicke Stakes). Springfield was linebred 5 x 4 to Sultan (son of Selim), 5 x 5 to Muley, and to the very close relatives Trampoline (dam of Glencoe) and Whisker (broodmare sire of Marsyas).

One further note of interest. Sir James Miller who bought Sainfoin as a three year old was a big fan of the mare Pocahontas. He told friends that he had worked out why Pocahontas was such a good producer. He said that in her pedigree: "Harpalice is in fact a close blood relation to Tramp's dam, both being by Gohanna from mares which trace to Everlasting by Eclipse. Furthermore, Harpalice is out of a mare by Trentham's son Driver, whereas Tramp's dam is out of Fraxinella, by Trentham." Upon investigating this statement, I found Sir James to be quite correct.

English Derby winner Manna (1922) has inbreeding to St. Simon and Bend Or and his pedigree is worth noting. Foaled in Ireland, Manna won 4 of 9 starts including the Richmond Stakes (at 2) and the English 2,000 Guineas.

Manna (1922)	Phalaris	Polymelus	Cyllene by Bona Vista, by *Bend Or*
			Maid Marian
		Bromus	Sainfoin
			Cheery by *St.Simon*
	Waffles	Buckwheat	Martagon by *Bend Or*
			Seasame by *St. Simon*
		Lady Mischief	*St. Simon*
			Vain Duchess

Manna's inbred dam Waffles was a temperamental mare yet she also produced Sandwich, winner of the English St. Leger. A peculiar thing about Manna's pedigree pattern is that he carries three daughters of unbeaten St. Simon, but no son. He sired the winners of 243 races and was leading sire of two year olds in England in 1933 and third on the sires' list the same year. Among his progeny was Colombo, brilliant winner of the English 2,000 Guineas.

Felstead, winner of the English Derby, is another inbred classic winner. He was the third Derby winner in male succession being by Spion Kop, by Spearmint. His pedigree is quite interesting showing a son and daughter of VRC Melbourne Cup winner Carbine (3 x 3) a stayer which was himself inbred to Brown Bess. Felstead's dam was inbred to Galopin:

Felstead (1925)	Spion Kop	Spearmint	*Carbine*
			Maid of the Mint
		Hamerkop	Gallinule
			Concussion
	Felkington	Lemberg	Cyllene
			Galicia by *Galopin*
		Comparison	William The Third by *St. Simon by Galopin*
			Combine by *Carbine*

Another English Derby winner with a classic pedigree pattern is Grand Parade, also winner of the St. James's Palace Stakes, Soltykoff Stakes, Angelsey Stakes and National Produce Stakes. His second dam Grand Marnier is full sister to Mother Siegel, dam of English Derby winner Minoru (an ancestor to be found in the pedigrees of Hyperion and Princequillo):

Grand Parade (1916)	Orby	Orme	Ormonde
			Angelica by Galopin
		Rhoda B.	Hanover
			Margerine
	Grand Geraldine	Desmond	*St. Simon by Galopin*
			L'Abbesse de Jouarre (by son of *Hermit)*
		Grand Marnier	Friar's Balsam by *Hermit*
			* mare by *Galopin*

* Grand Marnier's dam, the mare by Galopin, traces directly to Chanoinesse, full sister to Derby winner Hermit. Thus, Grand Geraldine is inbred twice to Hermit, plus his full sister. Notice how the most effective inbreedings involve both sexes, either sons or daughters, or full siblings (male and female).

Minoru is by Cyllene (second dam by Hermit) from Mother Siegel, sister to Grand Marnier. The Derby winner is therefore carrying Hermit 4 x 3, plus having Hermit's his full sister Chanoinesse. I find this quite intriguing. Hermit was an outstanding racehorse winning besides the Derby, 7 other races and was a stakes winner at 2 years. He broke a blood vessel a week before his Epsom victory and was a renowned sire of bleeders. However, his progeny were so superior on turf that breeders patronized him. Hermit sired Tristan (Ascot Gold Cup), triple crown filly Shotover (2,000 Guineas, Derby and St. Leger), Blaise (English Derby) and Heume (French Derby). The great stayer Tristan is of course the sire of English Oaks winner Canterbury Pilgrim, dam of Swynford and Chaucer.

A mare which became a successful producer for Mr. S. B. Joel in England was St. Josephine, by the inbred successful broodmare sire St. Denis. St Josephine produced a colt named Polyphontes, inbred three times to Galopin. Polyphontes won the Eclipse Stakes twice, Ascot Derby and was third in the English St. Leger. Please note that St. Denis is out of a mare by Blue Green whose dam Angelica (by Galopin) is full sister to St. Simon:

			Galopin
		St. Simon	*St.Angela*
	St. Denis		Blue Green (dam *Angelica, sister to St. Simon)*
		Brooch	Ornament
St. Josephine (1909)			Hampton
		Royal Hampton	Princess
	Nathalie		*Galopin*
		Nattie	Gnat

In 1919, a colt was born with intense inbreeding to the full brothers St. Hilaire (St. Gatien-Distant Shore) and St. Damien. The colt's name was Pondoland, also bred by Mr. S. B. Joel. This colt won the Newmarket Stakes, Sandown Anniversary Cup, Ascot Coventry Stakes, Berkshire Foal Stakes, Knowlsey Nursery, and was runner-up to Captain Cuttle in the 2,000 Guineas. Here is his pedigree:

			Cyllene ex Arcadia, ex *Distant Shore*
		Polymelus	Maid Marian by *Hampton*
	Pommern		*St. Hilaire by St. Simon-Distant Shore*
		Merry Agnes	Agnes Court by *Hampton*
Pondoland (1919)			*St. Simon*
		St. Damien	*Distant Shore*
	Gourouli		Minting
		Glave	Gloria

In this instance we find a stakes winner with a triple cross of the mare Distant Shore. His own sire Pommern is inbred to Distant Shore 4 x 3 and to Hampton 3x 3. Pommern won the English Derby, 2,000 Guineas, June Stakes, Richmond Stakes, Imperial Produce Stakes, September Stakes, and Limekiln Stakes so he was a tough little horse.

White Eagle, a decent broodmare sire (he is sire of the dam of Blandford, and also appears in the pedigree of Princequillo) is an inbred horse being by Gallinule (son of Isonomy) from Merry Gal by Galopin from Mary Seaton by Isonomy. He is therefore 2 x 3 to a son and daughter of Isonomy.

An English Derby winner with inbreeding to Isonomy was Lemberg (1907) by Cyllene (dam by Isonomy) from Galicia (whose dam was by Isonomy). Lemberg also won the St. James's Palace Stakes, Champion Stakes, Jockey Club Stakes, Lowther Stakes, Coronation Cup, Doncaster Cup, Middle Park Plate, Dewhurst Plate and was a champion, so his inbreeding was an advantage.

French Derby winner Herbager, sire of the dam of champion Rainbow Quest, is an outbred horse but his dam Flagette is inbred 2 x 2 to English St. Leger winner Firdaussi. Herbager also appears in the pedigree of U.S. champion 2YO colt Devil's Bag, and champion older mare Glorious Song, dam of Rahy.

Bold Promise, winner of the $1 million Magic Millions Stakes in Australia, bred by Mr. Ross Cribb, was a brilliant two year old inbred to the very fast English mare Arietta, dam of Ebor Handicap Gr-1 winner Sostenuto, and stakes winner Idomeneo. Arietta was by English champion Tudor Minstrel. Bold Promise also won the Riesling Slipper Trial Gr-2, Coca-Cola Classic Gr-3, Tea Rose Stakes Gr-2, and the Furious Stakes (Listed). She is dam of $2 million Golden Slipper Stakes Gr-1 winner Merlene which added the valuable AJC Sires' Produce Stakes Gr-1 to her record. Merlene is Australia's champion 2YO filly of 1996 and trained by Lee Freedman. Here is her dam's pedigree which does not show line-breedings to Hyperion and Lady Juror:

			Star Kingdom
		Kaoru Star	Kaoru
	Luskin Star		Idomeneo (ex *Arietta)*
		Promising	Modern Touch
Bold Promise			Bold Ruler
		Bold Aussie	Parida
	Game Dame		Sostenuto (ex *Arietta)*
		Royal Crest	Regal Peace

Her sire Luskin Star was the best 2YO ever to race "down-under" and zig-zags back to Arietta. I wonder did Bold Promise inherit Arietta's X chromosome via Luskin Star's dam? Incidentally, Merlene is by Danehill, already sire of three consecutive Golden Slipper Stakes Gr-1 winners. Here is his pedigree:

			Nearctic
		Northern Dancer	*Natalma*
	Danzig		Admiral's Voyage
		Pas de Nom	Petitioner
Danehill			Ribot
		His Majesty	Flower Bowl
	Razyana		Buckpasser
		Spring Adieu	*Natalma*

Danehill is leading sire in Australia and is a prepotent stallion. He won 4 races in England on turf including the Ladbroke Sprint Stakes Gr-1, Ascot Cork & Orrery Stakes Gr-3, European Free Handicap, and was third English 2,000 Guineas Gr-1 and July Cup Gr-1. He was champion 3YO sprinter in England. The 3 x 3 to Natalma has been most effective and a repeat mating supplied stakes winner Eagle Eyed, winner of the Arlington Classic Stakes Gr-2, Sausalito Handicap, Pirate Cove Stakes and now resident at Ashford Stud, Kentucky. Danehill was bred by Prince Khalid Abdullah's Juddmonte Farms. The Grade 1 miler Quiet American (Fappiano-Demure by Dr. Fager) which stands at Gainsborough Farm, is inbred 3 x 2 to champion Dr. Fager and inbred 4 x 3 to Cosquilla which traces in direct female line to Bull Dog's sister Marguerite de Valois. His pedigree fascinates me. In Quiet American's first crop he sired the outstanding filly Cara Rafaela and I am confident he will do well with mares carrying a "son" of Nearctic (e.g. Northern Dancer's strain).

Daniel Wildenstein's Epervier Bleu (Saint Cyrien-Equadif by Abdos) won the Prix Lupin Gr-1, Prix Greffulhe Gr-2, Grand Prix de Saint Cloud Gr-1, and Prix Niel Gr-2 a few years ago. He has a clever pedigree pattern involving duplication of champion Djebel four times (5 x 7 x 4 x 5) and his dam Equadif is inbred 2 x 3 to French mare Pretty Lady (by Umidwar).

Hydro Calido, a Group 2 winner out of the inbred mare Coup de Folie, has a 4 x 4 x 4 inbreeding to Almahmoud. Hydro Calido is by Nureyev and won the Prix d'Astarte Gr-2 for the late Stavros Niarchos. There are many Group winners with inbreeding or linebreeding to Mahmoud's famous daughter Almahmoud. It has been extremely effective.

Childwick, another good racehorse, winner of the Cambridgeshire Handicap and influential broodmare sire, is inbred 4 x 4 to Pocahontas and gets another source of Pocahontas via the damline of Chatanooga as follows:

			Vedette
		Galopin	Flying Duchess
	St. Simon		King Tom ex *Pocahontas*
		St. Angela	Adeline
Childwick (1890)			Chatanooga (*to Pocahontas)*
		Wellingtonia	Auraucaria ex *Pocahontas*
	Plaisanterie		Trocadero
		Poetess	La Dorette

He sired French Derby winner Negofol, and La Camargo, 23 wins and grand-dam of Massine, sire of French Derby winner Mieuxce. Childwick was a successful broodmare sire.

One of the superior sons of Solario was classic winner Mid-Day Sun which was the product of inbreeding to Sainfoin and Sierra, plus St. Simon and Hampton - a mix of speed and stamina. He won the English Derby, Hardwicke Stakes, Manton Stakes, Free Handicap, Derby Trial Stakes and was third in both the 2,000 Guineas and St. Leger. Here is a cleverly designed pedigree which was done on purpose by his breeder:

			Bayardo
		Gainsborough	Rosedrop by a son of *St. Simon*
	Solario		Sundridge (ex *Sierra, sister to Sainfoin)*
Mid-Day Sun		Sun Worship	Doctrine
			Polymelus
		Phalaris	Bromus by *Sainfoin, her dam by St.Simon*
	Bridge of Allan		Spearmint
		Spean Bridge	Santa Brigida by *St. Simon*

Irish Derby and Irish 2,000 Guineas winner Phideas is another with the successful Sainfoin and Sierra mix. He is by Pharos (Sainfoin comes via Phalaris's dam) out of Imagery, whose dam is by Sundridge (son of Sierra, sister to Sainfoin). Imagery of course is a full sister to St. Leger winner Solario, and is an important element in the pedigree of stakes winner Cox's Ridge.

Lord Derby's English Derby winner Sansovino (Swynford-Gondolette) is inbred 3 x 3 to classic winner Pilgrimage which is by The Palmer. Sansovino's pedigree pattern reveals The Palmer's brother Rosicrucian as the sire of Douranee, she being the third dam of Sansovino. Thus, this classic winner was inbred to full brothers three times, i.e. 4 x 4 x 4.

Raeburn. However, he sold the yearling colt to Mr. S. D. Hollingsworth. Raeburn won the Irish Derby, Column Produce Stakes and although 4 x 4 to Sierra, he is actually inbred 3 x 3 to brother and sister, Sundridge and Amphora:

			Bayardo
		Gainsborough	Rosedrop
	Solario		*Sundridge (Amphion-Sierra)*
		Sun Worship	Doctrine
Raeburn			John O'Gaunt
(1933)		Swynford	Canterbury Pilgrim
	Harpy		Flying Fox
		Flying Vixen	*Amphora (Amphion-Sierra)*

Harpy is to be found on Cigar's direct female line. In the above pedigree there is linebreeding to St. Simon. The examples of classic winners and stakes winners provided in this chapter show how effective reinforcement of superior strains can upgrade classic speed. Most of the examples given were successful in the breeding shed.

American breeder William Woodward created inbreeding to two unbeaten ancestors, St. Simon and Ajax, when he designed the match for The Scout II, winner of the Yorkshire Cup, Newmarket Stakes and Knowsley Dinner Stakes. Not shown in the pedigree which follows is the fact that Orme's dam Angelica is sister to St. Simon. This pedigree shows opposite sex strains, male and female:

			Ajax by Flying Fox by Orme
		Teddy	Rondeau
	Sir Gallahad III		Spearmint
		Plucky Liege	Concertina *by St. Simon*
The Scout II			*St. Simon*
		Rabelais	Satirical
	La Rablee		*Ajax by Flying Fox by Orme*
		La Flambee	Medeah

In the pedigree of Nearctic (sire of Northern Dancer) please try to appreciate the valuable genetic input of stakes winner Abbot's Trace. Let's first give Nearctic's pedigree and work backwards to show what I mean:

			Phalaris
		Pharos	Scapa Flow by *Chaucer (ex C. Pilgrim)*
	Nearco		Havresac II
		Nogara	Catnip by Spearmint
Nearctic			Gainsborough
		Hyperion	Selene by *Chaucer (ex C. Pilgrim)*
	Lady Angela		Abbot's Trace by Tracery
		Sister Sarah	Sarita by Swynford *(ex C. Pilgrim)*

Perhaps the first thing to note is the 5 x 5 x 5 linebreeding to Canterbury Pilgrim (via her sons Chaucer and Swynford). The second thing to note is a rather heavy concentration of St. Simon and his sister Angelica which doesn't show up in the short pedigree above. Nearco and Hyperion were both linebred to St. Simon as shown in another chapter. All duplicated ancestors in this match were top-class athletes.

Next we should investigate Sister Sarah, a mare from the Number 14 Family tracing to the peerless Pretty Polly, winner of the English 1,000 Guineas, Oaks, St. Leger, Middle Park Plate, National Breeders Produce Stakes, Champagne Stakes etc. Sister Sarah was half-sister to Lady Maureen third dam of Macherio, also third dam of French stakes winners Ocarina, Guersant and Flute Enchantee, dam of champion miler Luthier. Sister Sarah was half-sister to Golden Silence, third dam of St. Leger winner Premonition.

Furthermore, Sister Sarah was half-sister to Irish Derby winners Spike Island and Zodiac; and half-sister to Molly Adare, third dam of champion Brigadier Gerard, and grand-dam of Orthodox and Eleanor Cross. Irish broodmare sire Artaius has Molly Adare as his fifth dam. Thus, we find Nearctic's damline to be a high-class female line. Here is Sister Sarah's pedigree:

			Rock Sand
		Tracery	Topiary by Orme
	Abbot's Trace		Rightaway
		Abbot's Anne	Sister Lumley
Sister Sarah			John O'Gaunt
		Swynford	Canterbury Pilgrim
	Sarita		Desmond
		Molly Desmond	Pretty Polly

Let's take a look at Rightaway, broodmare sire of Abbot's Trace. Rightaway is an inbred horse, with 3 x 3 x 3 to Rataplan and his brother Stockwell:

			Rataplan (ex Pocahontas)
		Blinkhoolie	Queen Mary
	Wisdom		*Stockwell (ex Pocahontas)*
		Aline	Jeu d'Esprit
Rightaway			*Stockwell (ex Pocahontas)*
		Honiton	Flax
	Vanish		Orlando
		Retreat	Flight

Winner of the Clearwell Stakes, Rightaway was not a successful sire because he was never patronized by commercial breeders mainly because his sire Wisdom was renowned as a broodmare sire. Like Wisdom, Rightaway left behind some more than useful daughters which became stakes producers. Rightaway's pattern gives us a clue as to how he might breed winners. His zig-zag goes to his dam, thence to his broodmare sire Honiton, a not so accomplished son of Stockwell. The zig-zag continues from Honiton's dam to Flax. This is not the strongest part of the pedigree. Now let's take a look at Abbot's Trace. A chestnut horse foaled in 1917 by champion Tracery, Abbot's Trace won 6 of 23 starts including the Newmarket St. Leger, Derbyshire Plate, Union Jack Stakes, Liverpool Summer Cup and Stockton Handicap but he displayed speed at 2, winning the Michaelmas Stakes over 6 furlongs.

His sire Tracery was inbred to siblings St. Simon and Angelica, but more importantly, he carried the influence of Wellingtonia whose pedigree reveals two *daughters* of Pocahontas. In the pedigree of the dam of Abbot's Trace we find Stockwell three times and Rataplan once, making *four sources of sons* of Pocahontas. Hence, the mating comprises use of 5 foals* out of the *Double Copy* mare Pocahontas:

* King Tom (sire of the dam of St. Simon) son of Pocahontas
* Stockwell and Rataplan - sons of Pocahontas
* Araucaria and Ayacanoora - daughters of Pocahontas

			Springfield
		Sainfoin	Sanda
	Rock Sand		*St. Simon (to Pocahontas)*
		Roquebrune	St. Marguerite
	Tracery		Ormonde
		Orme	*Angelica (sister to St. Simon)*
	Topiary		Wellingtonia (*2 daughters of Pocahontas)*
		Plaisanterie	Poetess
Abbot's Trace			Blinkhoolie
		Wisdom	Aline by *Stockwell (ex Pocahontas)*
	Rightaway		Honiton by *Stockwell (ex Pocahontas)*
		Vanish	Retreat
	Abbot's Anne		Hermit
		St. Honorat	Devotion by *Stockwell (ex Pocahontas)*
	Sister Lumley		See Saw
		Lady Lumley	Lady Alice Hawthorn

Since Abbot's Trace has the same zig-zag line as Rightaway, he might be a similar sort of sire. That is exactly what happened! The best progeny he ever sired were not of the highest class, but were decent handicappers and include Abbot's Speed, 9 wins including the Jubilee Handicap (twice) which became the broodmare sire of Prince Chevalier; Sunny Trace, Marconigram - a stakes winner and leading broodmare sire in Australia - The Black Abbot and The MacNab, two big handicap winners.

When the mix of Abbot's Trace and Hyperion eventuated which created Lady Angela, the pedigree pattern was suddenly saturated with St. Simon, bringing in King Tom (son of Pocahontas). Lady Angela then was mated with Nearco, thus reinforcing King Tom and his dam Pocahontas. Thus, Nearctic, the result of this match, would become an influential sire for speed in Canada for Eddy Taylor's Windfields Farm. Apart from Northern Dancer, Nearctic also sired Icecapade. With so much St. Simon in his genotype, is it little wonder than Nearctic transmitted sickle shaped hocks, the characteristic so common in St. Simon's progeny?

Another interesting mating is that of Lemonora (1918), by Derby winner Lemberg from Honora by the soft-boned stallion Gallinule. Bred by Sir Mark Sykes and sold as a yearling to Lord Manton for 3,300 guineas, Lemonora won the Grand Prix de Paris, Doncaster Champagne Stakes (beating Humorist), Newmarket Stakes, was runner-up in the 2,000 Guineas (to Craig an Eran), and third in the English Derby (to Humorist, Craig an Eran):

			Bona Vista
		Cyllene	ARCADIA (*Isonomy-mare by Hermit)*
	Lemberg		Galopin
		Galicia	Isoletta by *Isonomy*
Lemonora (1918)			*Isonomy*
		GALLINULE	Moorhen by *Hermit*
	Honora		Saraband
		Word of Honour	Geheimniss

Lemonora is inbred 3 x 2 to Arcadia and Gallinule which are three-parts "genetic" sister and brother, being by Isonomy out of mares by Hermit.

English Triple crown winner Gay Crusader, perhaps a disappointment as a sire yet an influential broodmare sire, is 3 x 3 to Galopin via two daughter strains. Gay Crusader is broodmare sire of French champion Djebel, and of Gay Gamp in Ribot's pedigree. Next we come to one of the best sires in Japan during the 1980's, Northern Taste, now making his mark as a broodmare sire. His daughters have a genetic affinity with Halo's son Sunday Silence. Here is intense inbreeding to Hyperion's daughter Lady Angela:

			Nearco
		Nearctic	*Lady Angela (to Tracery)*
	Northern Dancer		Native Dancer
		Natalma	Almahmoud
Northern Taste			Chop Chop
		Victoria Park	Victoriana *(to Tracery)*
	Lady Victoria		*Hyperion*
		Lady Angela	*Sister Sarah (to Tracery)*

Remember, Tracery carries a strain of Wellingtonia, a stallion with two *daughters* of Pocahontas, namely Ayacanoora and Araucaria. The above pedigree shows 3 x 2 inbreeding to Lady Angela. Surely this would have increased prepotency? Please go back and re-look at Danehill's pedigree because he too, is inbred to an influential mare and became a leading sire. Might Danehill's full brother Eagle Eyed also inherit prepotency from the 3 x 3 inbreeding to Natalma? I don't know. It will depend upon probability factors - whether Eagle Eyed inherited similar genes to Danehill.

Moonshell (Sadler's Wells-Moon Cactus by Kris), winner of the 1995 English Oaks Gr-1 is linebred 4 x 5 to influential mare Lalun via her sons Bold Reason and Never Bend. Her pattern actually shows inbreeding because these sons of Lalun are more than half-brothers. Bold Reason is by Hail To Reason (grandson of Royal Charger) whereas Never Bend is by Nasrullah.

I trust these examples indicate the point that inbreeding can be an effective tool to shuffle genes around to "fix" certain desirable characteristics or traits so that they may be used as material for future generations. By reinforcing the best elements of a pedigree, particularly high-class mares, one is in fact manipulating the genotype, i.e. tinkering with "the genes". French Derby winner Bering (Arctic Tern-Beaune by Lyphard) is linebred to Almahmoud 3 x 5, and 5 x 5 to French stayer Prince Bio. He stood in Kentucky for several years and got Pennekamp (English 2,000 Guineas Gr-1), Matiara (French 1,000 Guineas Gr-1), Signe Divine (Criterium de Maisons Laffitte Gr-2, Prix Daphnis Gr-3), Serrant (Grand Prix d'Evry Gr-2) as well as Peter Davies (Royal Lodge Stakes Gr-1). Bering now stands at Alec Head's beautiful Haras du Quesnay near Deauville and I inspected him again on a recent visit. He is a magnificent stallion with excellent shoulder angle and plenty of bone. I would hope French breeders send him mares with Halo's influence, or test a repeat Northern Dancer to add a third source of Almahmoud. In time, we might see a Group 1 winner being bred this way. A doubling of Northern Dancer wouldn't hurt because Bering has plenty of forearm length and decent shaped hocks. Mares with Nureyev's strain might be an interesting test.

The final example I wish to offer is the inbred mare Sanctimony, a French mare which produced the high-class racehorse San Souci II. The second foal of his dam, Sans Souci II won over 2,000 metres as a juvenile in France in the Prix Destrier and was third in the Prix Futurity. At three, he won the Grand Prix de Paris, Prix Daru, Prix Lupin and traces on his damline to the *Double Copy* mare Pocahontas.

If the M-factor (mitochondrial DNA) is important, the influence from Pocahontas's direct female line might benefit Sans Souci's filly foals, but not his colts.

				Newminster (brother to Honeysuckle)
			Hermit	Seclusion
		Heume		Breadalbane (son of Blink Bonny)
			Bella	Armada
	Le Roi Soleil			Vermouth
			Boiard	La Bossue
		Mlle de la Valliere		Monarque
			Laversine	Voluptas by *Stockwell (ex Pocahontas)*
Sans Souci II				Galopin
			St. Simon	St. Angela by King Tom (*Pocahontas)*
		St. Serf		*Thormanby*
			Feronia	Woodbine *(by Stockwell-Honeysuckle)*
	Sanctimony			Doncaster
			Bend Or	Rouge Rose by *Thormanby*
		Golden Iris		Macaroni
			Gardenia	Araucaria *(ex Pocahontas)*

Sanctimony was inbred 3 x 4 to Thormanby (recessive transmittor) but more importantly carries three sources of Pocahontas via sons, plus one source via Araucaria, a daughter. Sans Souci II was twice leading sire in France. He also became a truly great broodmare sire in Europe. It is really amazing how a mare like Pocahontas keeps coming back all the time in Group winner's pedigrees.

NEARCTIC, sire of Northern Dancer and Icecapade.

PLANNING MATINGS FOR THE RACETRACK

" The test of a first-rate intelligence is the ability to hold two opposed ideas in the mind at the same time, and still retain the ability to function."
- F. Scott Fitzgerald

Private breeders have made more contribution to the improvement of the breed than have commercial breeders if one examines the great breeders of the 20th century. Whereas the commercial horse-breeder must make a profit to carry on with his trade, a wealthy person may own a big racing stable and never have to show a profit. Some wealthy owners with this attitude actually made a lot of money, particularly those in the United States of America.

However, there were other famous horse-breeders who had financial difficulties and experienced the hardship of wars during the early part of the 20th century. Such a person was Federico Tesio, the "wizard" of Dormello. Although he inherited money from his parents from a trust fund set up prior to their death, Tesio was basically orphaned as a child. For the first twenty years he made terrible mistakes and wasn't a force as a breeder or trainer in Italy. Travel was something he liked so he began to learn from experienced people in England and France and by the time he was forty, Tesio changed his way of *how to breed the racehorse.* He achieved everlasting fame as the breeder/owner of Nearco, Ribot and Donatello II, three great racehorses and successful sires. Tesio trained unbeaten Nearco and champion Italian three year old Donatello II.

Tesio was forced to sell horses in training to finance his breeding operation; sell breeding stock to Germans to buy horse-feed during the ravages of two World Wars; take on a rich partner that knew nothing about breeding racehorses; and suffer partial blindness towards the end when he needed his sight most of all. For in the end, he had a little horse called Ribot in his stable, and this horse was a "super" colt. Unfortunately, Tesio died before Ribot ever made his appearance on the racetrack. It was left to the wealthy partner Marchisa della Rochetta (and Tesio's wife Donna Lydia) to race unbeaten Ribot and win two Prix de l'Arc de Triomphes.

Federico Tesio was a genius with horses. He didn't worry about fashion. He bred horses the way he wanted to breed them, irrespective of criticism from his friends and other trainers. For his time, he was the greatest horse-breeder in the world, with a very mediocre budget that had to be stretched to make ends meet.

Tesio's idea of breeding stayers would be scoffed at by modern breeders. Now, everyone wants speed! In my first book called "Tesio - Master of Matings" I outlined the talents of this famous trainer, owner and horse-breeder. I attempted to show how Tesio created matings to develop a fast stayer like Nearco. He used the techniques of inbreeding, linebreeding and outcrossing - whatever he felt was "the right thing to do". He tried them all. Most of his success came from matings designed to duplicate St. Simon and his sister Angelica, and pedigree patterns involving sons and daughters of Cyllene.

Today, a horse-breeder with a limited budget must set realistic objectives and start with a few quality mares and try to upgrade them for the racing stable. The name of the game is quality, not quantity. Mares to be secured *must* be above average for anyone to have a chance to succeed.

Unlike a commercial breeder who has to take his products to market, the private breeder can use stallions which were Grade 1 or Group 1 winners which are affordable because the commercial breeders decide to bypass them for some reason or another. To give an example, here is a list of really good racehorses standing in Kentucky at very reasonable fees:

SUNNY'S HALO (Halo-Mostly Sunny by Sunny, son of Princequillo)
Stands at Walmac Farm for $5,000 live foal. Champion 2YO colt Canada, won Kentucky Derby Gr-1, Super Derby Gr-1, Arkansas Derby Gr-1, Coronation Futurity Gr-1, Grey Stakes Gr-2 etc. Earned $1.2 million. Sound racehorse. Good shoulder angle. Sire of 20 stakes winners including Dispersal ($1.5 million earner), Sunny Sunrise ($1.2 million earner), Irgun ($475,600 1st Wood Memorial Stakes Gr-1), Race The Wild Wind ($458,250 earner 1st Santa Maria Hcp Gr-1) etc.
He needs mares with: *Northern Dancer, Princequillo, Nasrullah, His Majesty.*

CALLER I.D. (Phone Trick-Plagiarizing by Ramsinga)
Stands at Pin Oak Farm for $6,000 live foal. Third best 2YO of his year, 4 wins from 8 starts at 2, $380, 958 earner, won Arlington Washington Futurity Gr-2 (easily), Saratoga Special Stakes Gr-2 by 9 lengths, Sanford Stakes Gr-3 by 5 1/4 lengths, 3rd Hopeful Stakes Gr-1 etc. Injured rear left leg in accident in Florida as an early 3YO which ended a promising career. Linebred to Nasrullah and his full sister Malindi, plus their 3/4 brother Royal Charger - thus duplicating champion Mumtaz Mahal. Has 2YO (first crop) in 1996.
He needs mares with: *Mahmoud, Turn-To, Ribot, Mr. Prospector, Buckpasser.*

DEMONS BEGONE (Elocutionist-Rowdy Angel by Halo)
Stands at Claiborne Farm for $5,000 live foal. Ran as Kentucky Derby favorite. Half-brother to Preakness Stakes Gr-1 winner Pine Bluff. Won 7 of 14 starts for $609,944 1st Arkansas Derby Gr-1, Rebel Stakes, Southwest Stakes, Ahoy Stakes, 2nd Razorback Handicap Gr-2. By a Preakness Stakes Gr-1 winner. Linebred to Mumtaz Mahal. Dam by classic sire Halo. Sire of Danville ($444,810 earner), Flitch ($388,941 earner), Beavers Nose, Shoal Creek, A Demon a Day etc.
He needs mares with: *Northern Dancer, Mahmoud, Wavering Monarch, Icecapade.*

These three stallions standing for $6,000 or less, all with something to offer the man or woman who breeds to race, are just examples. There are many other stallions with the same genetic qualities as the above at reasonable fees. What one has to do is judge the value of stallions because some are under-priced and represent terrific value. Others could be way over-priced and outside your budget. To give another example, a stallion like Known Fact gets stakes runners, yet isn't as popular as he should be with commercial breeders even though he has sired European champion Warning. He is a proven sire which could be selected and used for a racing stable's program. He stands for only $8,000 fee at Juddmonte Farms. Wavering Monarch, sire of two champion juveniles, is another that represents tremendous value. He stands at Glencrest Farm. In 1996 his fee was $7,500 live foal. One has to shop around. Wavering Monarch's dam is by champion Buckpasser. These stallions have the potential to sire really high-class Grade 1 and Group 1 winners if they can cover the "right" mares. Gilded Time, Dehere, Brocco and Maria's Mon are four American champion juveniles with great sire prospects. I advocate using 2YO champions (i.e. young unproven sires) because I feel they are a worthwhile risk. In Europe, Wolfhound looks a real chance at a reasonable fee. Arazi's first crop

impressed me. However, In Ireland CATRAIL (Storm Cat-Tough as Nails) should be the big surprise for he is bred along similar lines to Danehill with a similar mix of Ribot, Buckpasser and Northern Dancer strains, and a doubling of elite mare Almahmoud (dam of Natalma). Australia's leading sire Danehill is inbred 3 x 3 to Almahmoud's best daughter, Natalma.

The breeder of homebreds must use skill when mare matings are planned. He or she must study the pedigrees of Group and Graded stakes winners and spot any new genetic affinity that can be copied. Selection of sound racehorses as sire material is crucial to ensure your trainer won't complain when homebreds are being conditioned at the track. Decide what each mare needs to upgrade her future foals. If she is bred along staying lines then choose a horse which matured quickly - one possessing plenty of fast-twitch muscle fiber. Please inspect stallions on your short-list, observing their temperament, conformation (including width of jaw) and character. The ones selected should supply conformation strengths not apparent in your mares. Try to visualize whether a stallion can actually supply clean airflow, correct shoulder angle and enough bone.

Keep in mind the stallion's dam, for she will be important if you get a filly foal. Temperament is something which must always be improved. Find out as much as possible about the stallion's temperament - check up on the temperament of his foals. Brilliant race fillies probably require a degree of stamina in their mating partners to ensure the foal will get at least a mile. Be reluctant to put pure speed to pure speed - it might cause the foal's shoulder angle to become too upright. Fast fillies require partners which could go around two turns (at least 1 1/8 miles). Speak with your trainer and ask about your mare's previous foals - learn about weaknesses being transmitted, and try to compensate with the mating planned for next season. Gathering information can help in the decision-making process. If you like a certain sire which is slightly beyond your budget, talk with the farm manager where the stallion is based and see if an *alternate foal share* might be arranged whereby you get one foal, and they get one foal, by the selected stallion. A written agreement is all that is required. If your mare comes from a stakes winning family, chances are the farm could be really interested. The conformation of the modern racehorse, its speed, its stamina, are the result of breeding from stallions and mares which carry in their ancestry many outstanding performers on dirt and/or turf. Selection has been made with above average racing merit as the goal, rather than

MOSTLY SUNNY, dam of Canadian champion 2YO Sunny's Halo, winner of the Kentucky Derby Gr-1.

physical type. The "selection process" has not been programmed to enhance fertility, physical beauty, nor perfect air-flow. *Soundness has always been a secondary goal.* Commercial breeders will use a fast stakes winner even if it failed to train on as a three year old. (There are of course genuine exceptions when a colt is injured due to accident). Private breeders, breeding for racetrack performance, are able to set different goals to those of commercial breeders. Utilization of very sound stallions becomes a necessity. Soundness *plus* speed are emphatic objectives to be set for one's breeding and racing program and these two goals can be achieved in conjunction with stamina. Whereas most commercial stallions possess acceptable conformation, and degrees of soundness, it has been the habit of thoroughbred breeders to be not as critical concerning fillies and mares which are used indescriminantly for breeding purposes. Unsound fillies are difficult to upgrade. They can pass on traits like knee chips, sesamoid weakness, bad feet, tendon weakness, left larygeal paralysis etc. Many young mares are "poor" breeders possessing sub-normal cycles inherited from their female ancestry and regularly fail to conceive. These mares should be eliminated from your breeding stock portfolio and replaced with mares of quality. Small stallions and mares are not at a disadvantage on the racetrack if they possess symmetry, strength through the hindquarters, clean respiratory systems, ideal circulatory systems and large hearts. Most small racehorses lack height because of shortness of forearms and gaskins. However, stakes winning horses of 15.0 to 15.2 hands which possess bodies relatively nearer to 16 hands can compete on an equal basis. We've seen lots of game, genuine little racehorses such as Rabelais, Hyperion, Selene and Star Kingdom (names from the past) and more recently Northern Dancer, Riverman, Blushing Groom and Lyphard.

To develop a successful racing stable, you have to be smarter than your competitor when it comes to planning matings. Your competitor will plan matings based on selecting a suitable sire for each mare (the foal's parents) just like a commercial breeder. Be smarter, go back one more generation of the proposed mating and look at the future foal's grandparents. Great breeders like Tesio, Lord Derby and Prince Aly Khan *always focused on the four parts of any proposed mating - each grandparent.* The four grandparents of any desirable mating are a breeder's best guide to guess what is likely to be transmitted from the consorting of stallion with mare. The grandparents are the custodians of the four major genetic influences in the foal's genotype and are "carriers" of four sets of genetic codes which can be shuffled around, mixing dominant and recessive traits in a pool - the gene pool. (Please go back to the book's introduction and re-read Mendel's pea experiments).

What a foal *inherits* from its *grandparents* is the *determining element* of whether you get a stakes runner or an allowance runner. From forty years experience, I now know that one must *concentrate on lining up performance in the grandparents* of a foal to maximize performance. If *all* four grandparents of a proposed pedigree are well above average, then the future foal has a better chance to become a stakes runner. If only two grandparents in a proposed pedigree have above average performance (the other two with poor performance) then the future foal is likely to be below average. Performance of grandparents is a key factor to consider when planning matings for the racetrack. Probability factors always affect a mating, no matter how clever it may be in the way it is constructed with duplicated superior ancestors, but to be ultra successful one must *purposely design* pedigree matches incorporating *outstanding grandparents* if you want a chance to breed stakes performers. From my observation, the majority of foals born each year take more of their phenotype (visual appearance) from one of their grandparents rather than from a parent. I've seen a foal look exactly like one of its ancestors in the second or third generation (remove) and we use the term "throw-back" to decribe this phenomenon. The foal very much resembles its more distant ancestor. Champion racehorses which disappointed as sires of stakes winners often have their influence expressed in their grandsons or grand-daughters. It is a paradox that triple crown winners were better broodmare sires than they were as sires of winners.

MYTHS AND MISCONCEPTIONS

" Discovery consists of observing what everybody has seen, and thinking what nobody has thought before."

" It's no good using that stallion now, he's about 25 years old," was a comment I overheard about Mr. Prospector at last year's sales. I turned around to see who the idiot was that made that comment. Obviously the guy didn't know Mr. Prospector is in excellent health, is the greatest living sire in North America with the winners of almost $64 million, including 142 stakes winners and 14 champions. His sons at stud include Gone West, Machiavellian, Gulch, Conquistador Cielo, Forty Niner, Rhythm, Distant View, Lion Cavern, Seeking the Gold, Kingmambo, Miswaki and Woodman.

Why wouldn't you want to breed to Mr. Prospector with a record like that!

He has always been a fertile stallion, his yearlings fetch big money, and his turf runners earned $12.6 million of his lifetime progeny earnings. His fee might not be in everyone's range, but he is still the best value stallion in Kentucky. Age has nothing to do with a stallion's ability to continue to sire stakes winners. It is not age that should be considered. When a stallion reaches more than 21 years of age, one has to consider his fertility and health. His genes are fixed in his genotype and don't change just because he is older: only his *quality of sperm* could be affected by ill health, a virus, or poor feeding/environment. Thus, an older stallion in good health at a farm like Claiborne should still manufacture quality sperm. Hopefully Mr. Prospector will be covering mares for a few more years.

The racing world is a strange place. It's human nature I suppose, for people needing to gossip. When I was a young lad I remember hearing "tales" about trainers, jockeys and racehorses. Myths magically manage to survive, even in modern times.

"Please don't buy a horse with four white feet." Ever heard this expression? It is misconception handed down from old trainers to young ones and refers to horses with markings on all four pasterns, or white running up all four cannon bones. Sure, some horses inherit those undesirable "soft" white hooves which bothers trainers and farriers and causes them to complain: yet, I've come across many Group/Graded stakes winners with four white hooves that were hard and tough like those of The Minstrel (champion 3YO colt of England) and Bering (French Derby winner, 2nd Prix de l'Arc de Triomphe Gr-1). One has to look at the type of hoof and see if it is strong or weak.

"Don't buy a colt from an old mare." - If enough people agree with this myth, then perhaps we'll have more bargains at sales. Today, many horse-breeders lease "nursemares" to raise foals from mares older than 20 years of age. This helps to ensure foals out of old mares receive more than enough nourishment while their real dam rests. Nursemares are a blessing to our breeding industry. What would we do without them? Too many high-class racehorses were foals of mares 16 years and over. Bois Roussel, winner of the English Derby, was born when his dam was 23 years old. Ksar, winner of two Prix de l'Arc de Triomphes was foaled when his dam was nearly nineteen. Health is paramount for mares of any age, not just old mares. Genes don't change with age. Good health is a always a vital factor in a broodmare - bad health will affect the quality of an older mare's ovum.

"He's a chestnut, and all the best runners by this sire are bays or browns." - Does one really, honestly think that color makes a horse run faster? Come on, think about this statement with some

degree of seriousness. Of course it doesn't. Color is inherited by one or two genes affecting the pigmentation of a horse's coat. The cells of the coat are shaped so as to reflect different colors, such as brown, black, bay, chestnut or dun. Gray is not a color. Horses with gray coats inherit a gene that causes a gradual loss of pigmentation.

In North America, it is common practice to describe a chestnut foal which inherits the gray *modifying gene* as a *"roan"*. Genetically, it is a chestnut with a modified gene, no matter how the roan appearance changes. A black, bay or brown foal which inherits the gray *modifying gene* will look a steel color of various shades, maybe with splashes of white ticked throughout its body until it gradually loses all pigmentation and becomes an almost white-looking animal.

Horses which have no color at all come in two separate classifications. A pure white horse with dark eyes is a recessive white, with no pigmentation of coat color. In this instance, the thoroughbred does not inherit a *gene* for color pigmentation - it lacks a few chemicals in its genetic code. However, there is another type of white racehorse, very rare, but still present within the racehorse gene pool. I refer to the "albino", with pink or clear eyes which suffers from blindness during sunlight hours but is extremely active at night, with excellent vision.

In conclusion, I'm afraid there is absolutely no connection between color and racing ability. It is purely coincidence or "chance" that things happen to encourage people to believe in fairytales about color.

"Don't buy a pidgeon-toed yearling." - This remark refers to colts or fillies which toe-inwards with both front feet. Personally, I consider yearlings which toe-in slightly to be a sign of speed. If a toe-in is so severe as to be unacceptable, then one must consider it a handicap and a fault - take it off your short-list. However, if a brood mare toes-in, it doesn't generally bother me so much because her racing days have ended, and it is a recessive trait, anyway.

Myths and misconceptions occur in the thoroughbred industry. I am not about to list them all. By far the majority of myths are groundless and often hinder progress. Sometimes we see horses with black tickings scattered throughout their coat. It is a legacy from an ancestor called Birdcatcher, a chestnut horse foaled in Ireland in 1833. He was by Sir Hercules from Guiccioli, a mare by Bob Booty, and I suspect the gene for it was inherited from Bob Booty's line. Birdcatcher (often referred to as Irish Birdcatcher) won 7 of 15 starts and was the father of The Baron, sire of Stockwell and Rataplan which perpetuated the black tickings via their many descendants. It was a myth in Ireland that yearlings with "Birdcatcher's tickings" would make good racehorses, but the truth of the matter is that many outstanding runners had Stockwell and/or Rataplan fairly close up in their pedigrees.

There is a misconception about dirt versus turf runners. Certainly the size and shape of the feet of a racehorse can help or harm it on different racing surfaces. Mr. Prospector has small feet, ideal for dirt performance. Progeny with smallish feet perhaps didn't handle turf as well as they might, especially on soft ground. Generally speaking, the reason why specific horses cannot or will not perform on both surfaces is because of structural or physical weakness. Racehorses with long pasterns or weak pasterns are at an absolute disadvantage on dirt; but then again, America's racetracks vary so much that it is perhaps dangerous to generalize.

Cigar was trained on dirt at the beginning of his racing career. He was at that time unspectacular. You can check his race record. He was then moved up to turf and won races. Later, Bill Mott decided to re-test him on dirt and Cigar became a sensation. Every horse is an individual and only by testing each individually on different surfaces will one ever know if they can handle one, or both.

Classic speed is the essence of victory. To make a comparison between racehorses of the past with stakes winners of recent times is not unfair if one takes into consideration four factors: (1) The improvement of racing surfaces and type of track - undulating or flat. (2) The different style jockeys now use, especially their front-running tactics. (3) How "tight" or "close" jockeys ride their mounts in races (compare very tight race riding in Australia with that of North America), and (4) how times are recorded, i.e. hand-timed versus electronically timed, and when the time begins, which varies in different countries.

COMPARISON OF TIMES

There are misconceptions about how much improvement has been made in the past fifty years of thoroughbred racing. Allow me to show by example, times recorded for classics run on turf and dirt. The classics selected were: the English Derby, 1 1/2 miles on turf at Epsom race course which is a left-handed undulating track with a downhill run to Tattenham corner, and rising ground to the winning post; The English Oaks (restricted to three year old fillies), 1 1/2 miles on turf also at Epsom race course; The English St. Leger, 1 3/4 miles at Doncaster, which is an undulating turf course; the Kentucky Derby, 1 1/4 miles on dirt around a flat-surfaced track; and to take older horses into consideration, the Prix de l'Arc de Triomphe 2,400m at Longchamp.

The purpose of providing these times is to make younger people aware of the talented horses of past years which ran fast times. Subjective judgement on my part was necessary to restrict the list for a meaningful summation by only supplying a selection of representatives, but I tried to be as fair as possible. There is no point accumulating figures if it is done with bias. Everyone knows that figures can be made to lie. Please consider inclement weather when assessing some of the slow times found on the list.

ENGLISH DERBY 1 1/2 miles turf (2,414 metres since 1872)

1850	Voltiguer	2m 50 seconds
1864	Blair Athol	2m 43
1867	Hermit	2m 52
1886	Ormonde	2m 45.6
1890	Sainfoin	2m 49.8
1896	Persimmon	2m 42
1903	Rock Sand	2m 42.6
1905	Cicero	2m 39.4
1906	Spearmint	2m 36.8
1914	Durbar II	2m 38.4
1920	Spion Kop	2m 34.8
1930	Blenheim	2m 38.2
1933	Hyperion	2m 34.4
1936	Mahmoud	2m 33.8 (record)
1937	Mid-day Sun	2m 37.6
1938	Bois Roussel	2m 39.2
1950	Galcador	2m 36.8
1957	Crepello	2m 35.4
1960	St. Paddy	2m 35.8
1970	Nijinsky II	2m 34.68

1971	Mill Reef	2m 37.14
1982	Golden Fleece	2m 34.27
1984	Secreto	2m 39.12
1985	Slip Anchor	2m 36.23
1986	Sharastani	2m 37.13
1990	Quest For Fame	2m 37.26
1991	Generous	2m 34.00
1993	Commander in Chief	2m 34.15
1996	Shaamit	2m 35.05

ENGLISH OAKS 1 1/2 miles turf (2,414 metres since 1872) 3YO fillies

1887	Reve d'Or	2m 50.6
1890	Memoir	2m 40.7
1893	Mrs Butterwick	2m 44.8
1896	Canterbury Pilgrim	2m 45.6
1901	Cap and Bells	2m 44.4
1902	Sceptre	2m 46.6
1905	Cherry Lass	2m 38.0
1927	Beam	2m 34.6
1933	Chatelaine	2m 34.6
1938	Rockfel	2m 37.4
1949	Musidora	2m 40.2
1950	Asmena	2m 42.6
1956	Sicarelle	2m 42.0
1959	Petite Etoile	2m 35.8
1960	Never Too Late II	2m 39.2
1970	Lupe	2m 41.46
1971	Altesse Royale	2m 36.95
1982	Time Charter	2m 34.21
1984	Circus Plume	2m 38.96
1985	Oh So Sharp	2m 41.37
1986	Colorspin	2m 35.60
1991	Jet Ski Lady	2m 37.30
1993	Intrepidity	2m 34.19
1996	Lady Carla	2m 35.55

ENGLISH ST. LEGER 1 3/4 miles turf

1886	Ormonde	3m 21.4
1890	Memoir	3m 13.6
1892	La Fleche	3m 14.6
1893	Isinglass	3m 13.4
1900	Diamond Jubilee	3m 09.2

English St. Leger (Continued)

1903	Rock Sand	3m 09.4
1904	Pretty Polly	3m 05.8
1906	Troutbeck	3m 04.2
1926	Coronach	3m 01.6
1933	Hyperion	3m 06.8
1937	Chulmleigh	3m 07.2
1938	Scottish Union	3m 11.6
1950	Scratch	3m 09.0
1957	Ballymoss	3m 15.6
1959	Cantelo	3m 04.6
1960	St. Paddy	3m 13.2
1961	Aurelius	3m 06.6
1969	Intermezzo	3m 11.45
1970	Nijinsky II	3m 06.40
1974	Bustino	3m 09.02
1975	Bruni	3m 05.31
1990	Snurge	3m 08.78
1991	Toulon	3m 03.12
1993	Bob's Return	3m 07.85

KENTUCKY DERBY 1 1/4 miles dirt

1896	Ben Brush	2m 07.75
1911	Meridian	2m 05.00
1915	Regret	2m 05.75
1923	Zev	2m 05.40
1926	Bubbling Over	2m 03.80
1931	Twenty Grand	2m 01.80
1937	War Admiral	2m 03.20
1941	Whirlaway	2m 01.40
1948	Citation	2m 05.40
1953	Dark Star	2m 02.00
1955	Swaps	2m 01.80
1958	Tim Tam	2m 05.00
1962	Decidedly	2m 00.40
1964	Northern Dancer	2m 00.00
1965	Proud Clarion	2m 00.60
1969	Majestic Prince	2m 01.80
1973	Secretariat	1m 59.40 (record)
1977	Seattle Slew	2m 02.20
1978	Affirmed	2m 01.20
1981	Pleasant Colony	2m 02.00

1983	Sunny's Halo	2m 02.2
1985	Spend A Buck	2m 00.2
1988	Winning Colors	2m 02.2
1989	Sunday Silence	2m.05.0
1990	Unbridled	2m 02.0
1993	Sea Hero	2m 02.8
1994	Go For Gin	2m 03.8
1995	Thunder Gulch	2m 01.2
1996	Grindstone	2m 01.0

PRIX DE L'ARC DE TRIOMPHE 2,400m turf (3YO and up)

1920	Comrade	2m 39.00
1921	Ksar	2m 34.80
1924	Massine	2m 40.80
1928	Kantar	2m 38.80
1931	Pearl Cap	2m 39.00
1934	Brantome	2m 41.75
1937	Corrida	2m 33.88
1942	Djebel	2m 37.80
1948	Migoli	2m 31.60
1949	Coronation V	2m 33.22
1951	Tantieme	2m 32.84
1956	Ribot	2m 34.76
1963	Exbury	2m 34.98
1965	Sea Bird II	2m 35.52
1969	Levmoss	2m 29.00
1971	Mill Reef	2m 28.30
1972	San San	2m 28.00
1977	Alleged	2m 30.60
1979	Three Troikas	2m 28.90
1980	Detroit	2m 28.00
1987	Trempolino	2m 26.30
1990	Saumarez	2m 29.80
1992	Suave Dancer	2m 31.40
1994	Carnegie	2m 31.10

Considerable improvement of times were accomplished when American jockeys invaded Britain earlier this century. Their different tactics of riding close to the pace, together with short stirrups, made an impact on times. The reader may judge whether much improvement was made in the past fifty years. Certainly in the 1970's the Prix de l'Arc de Triomphe established faster times, and ironically, three fast times set by fillies. Secretariat's sub-two minute time for the Kentucky Derby is yet to be bettered.

It is a myth that one needs big horses to win big races. Size in thoroughbreds is generally between 15 to 17 hands high, and many of the 17 hands horses have difficulty at the starting gate, are slow to get into stride, and suffer more respiratory trouble than do small horses. One hand is the equivalent of 4 inches.

For interest sake, I list numerous stallions and mares, detailing their height:

Stallions		*Hands High*	Mares		*Hands High*
Second half of 18th Century:					
Matchem	**(1748)**	**15.1**	**Queen Mab**	**(1785)**	**14.3**
Herod	**(1758)**	**15.3**			
Eclipse	**(1764)**	**15.2**			
Sir Peter	**(1784)**	**16.0**			
Gohanna	**(1790)**	**15.1**			
Sorceror	**(1790)**	**16.1 1/2**			
Orville	**(1799)**	**16.0**			
Middle of 19th Century:					
Touchstone	**(1831)**	**15.2**	**Velocipede's dam**	**(1817)**	**15.0**
Birdcatcher	**(1832)**	**15.3**	**Beeswing**	**(1833)**	**15.2**
Melbourne	**(1834)**	**16.0**	**Pocahontas**	**(1837)**	**14.3**
Newminster	**(1842)**	**15.2**	**Miami**	**(1844)**	**15.1**
Stockwell	**(1849)**	**16.2**	**Blink Bonny**	**(1854)**	**15.2 1/2**
Lexington	**(1850)**	**15.3**			
Macaroni	**(1860)**	**15.3**			
Blair Athol	**(1861)**	**16.0**			
Late 19th Century:					
Springfield	**(1873)**	**16.0**	**Sanda**	**(1878)**	**15.3**
Isonomy	**(1875)**	**16.0**	**Atalanta**	**(1878)**	**15.1**
Bend Or	**(1877)**	**16.1**	**Perdita II**	**(1881)**	**15.1**
St. Simon	**(1881)**	**16.0 1/2**	**Morganette**	**(1884)**	**16.1**
Ormonde	**(1883)**	**16.1**	**Ornament**	**(1887)**	**16.0**
Carbine	**(1885)**	**15.3 1/2**	**La Fleche**	**(1889)**	**16.0**
Sainfoin	**(1887)**	**15.2 1/2**	**Medora**	**(1890)**	**15.2**
Orme	**(1889)**	**16.1**	**Festa**	**(1893)**	**15.2**
Childwick	**(1890)**	**16.0**			
Isinglass	**(1890)**	**16.1**			
Persimmon	**(1893)**	**16.2**			
Flying Fox	**(1896)**	**16.0 1/2**			
First half of 20th Century:					
Chaucer	**(1900)**	**15.2 1/2**	**Selene**	**(1919)**	**15.1**
Phalaris	**(1913)**	**16.1**	**Plucky Liege**	**(1912)**	**15.3**
Hurry On	**(1913)**	**17.0**	**Nogara**	**(1928)**	**15.3**
Blandford	**(1919)**	**16.0**	**Lavendula II**	**(1935)**	**16.0**
Hyperion	**(1930)**	**15.2**	**Rockfel**	**(1935)**	**16.0**
Donatello II	**(1934)**	**16.1**	**Flaming Top**	**(1941)**	**16.2**
Fair Trial	**(1932)**	**16.0**			
Nearco	**(1935)**	**16.0**			

Nasrullah	**(1940)**	**16.1**
Star Kingdom	**(1946)**	**15.2**
Tantieme	**(1947)**	**16.3**

Second half of 20th Century:

Hugh Lupus	**(1952)**	**16.2**
Ribot	**(1952)**	**15.3**
Northern Dancer	**(1961)**	**15.2 1/2**
Bold Bidder	**(1962)**	**16.2**
Grey Dawn II	**(1962)**	**16.3**
Tim Tam	**(1955)**	**16.1 1/2**
Caro	**(1967)**	**16.3**
Nijinsky II	**(1967)**	**16.1**
Lyphard	**(1969)**	**15.2**
Halo	**(1969)**	**16.0**
Sec retariat	**(1970)**	**16.0**
Mr. Prospector	**(1970)**	**16.0**
Blushing Groom	**(1974)**	**15.2**
Storm Bird	**(1978)**	**16.0**
Never Too Late II	**(1957)**	**15.3**
Three Troikas	**(1976)**	**16.1**
Terlingua	**(1976)**	**16.0**
Mrs. Penny	**(1979)**	**16.0**
Height of Fashion	**(1979)**	**16.3**
Miesque	**(1984)**	**16.0 1/2**

The size of thoroughbreds improved by 1900 and hasn't changed much. However, from the list, it is perhaps surprising that Pocahontas, Perdita II, Atalanta and Selene were quite small mares.

There exists misconception about "roarers", some of which transmit the trait to a small proportion of foals, while others do not. What is confusing is that a horse like Ormonde, a triple crown winner, did not exhibit any noise while in training as a two or three year old, at least not according to lore. At 4, he could be heard making loud noise in work, so whether it was congenital or an environmental cause, one can't be certain. One shouldn't assume too much. Notable among horses which were "roarers" were Pocahontas, Ormonde, Chamant, Gallinule, Coronach, Humphrey Clinker, Rayon d'Or and Prince Charlie (according to the history books). Domino and Bend Or seem to have had trouble with their left larynx, because they transmitted left paralysis to a tiny proportion of their foals. According to The Bloodstock Breeders' Review (1937 edition, pages 145-147) : "Laryngeal operations for the relief of roaring were first recorded in 1834. A German veterinary surgeon Herr Gunther discovered that roaring and whistling were due to paralysis of the recurrent nerve or nerves functioning the muscles which abduct the artenoid cartilages of the larynx during inspiration. For the relief of the lesion, he removed one or both of the sunken artenoid cartilages together with one or both vocal chords. Many modifications of Gunther's operation (some widely different) were subsequently tried in England and on the Continent before and after either the American Professor Williams or Sir Frederick Hobday entered the arena."

Exhaustive studies on the inheritance of roaring in various countries indicate that the problem is inherited as a recessive factor. This means that sound-winded stallions and mares may carry the trait in a state of latency and that it will be unknowingly present in some of their descendants. Stockwell and King Tom never inherited a problem from Pocahontas, their dam, nor is it likely they transmitted it to the first generation of their progeny.

However, there is suspicion that Rataplan (Stockwell's brother) may have inherited the recessive gene, because a few of his foals were roarers. Whether this problem was inherited from Rataplan, or via the dams of those foals, one will never really know.

BLUSHING GROOM'S FAMILY

"Talent develops in quiet places, character in the full current of human life".

- Johann Wolfgang von Goethe

Among the foals born in the northern hemisphere during 1954 was a small chestnut colt with a genuine attitude. By the dominant Nasrullah, he was out of an American-bred mare named Spring Run, daughter of Tom Fool's sire Menow. His name was Red God.

A stakes winner of 5 races from 2 to 4 years in England and the USA, Red God won the Richmond Stakes, the Roseben Handicap, and was runner-up in both the Champagne Stakes and Brandywine Turf Handicap to earn $56,565. Other sons of Nasrullah earned heaps more, like Nashua, Bold Ruler, Never Say Die and Noor. Upon retirement, the little chestnut was sent to Ireland. To the surprise of nobody, he began to sire winners. A lot was expected from sons of Nasrullah at the time. He attracted better mares and became a popular sire of yearlings, especially after the appearance of Folle Rousse, 14 wins, champion 2YO filly in England, 1st Acorn Stakes, Prix Robert Papin, King Neptune Handicap, and the very smart Jacinth, 5 wins, champion 2YO filly in England, 1st Coronation Stakes Gr-2, Falmouth Stakes Gr-3, 2nd English 1,000 Guineas Gr-1.

In due course he sired Green God (6 wins, champion sprinter in England), Greenland Park (Queen Mary Stakes Gr-2, Molecomb Stakes Gr-3), Red Lord (French 2,000 Guineas Gr-1), Yellow God (Gimcrack Stakes, Prix du Palais Royal) and Soul Mate (Correction Handicap); but when a small chestnut colt named Blushing Groom appeared in France in 1976, Red God became "hot". Here was a special colt that looked top drawer.

Foaled in France in 1974, Blushing Groom (Red God-Runaway Bride by Wild Risk) was bred by Mr. J. McNamee Sullivan, an Irish gentleman who acquired the colt's dam from the Aga Khan's farm. Runaway Bride was a plainish type with no form whatsoever, having placed only once at 4 years, but she had been culled from one of the world's most successful breeding operations. Prior to selling Runaway Bride, His Highness Karim Aga Khan had bred from her a colt named Bayraan (by Red God) which won 5 races including the Prix de Seine-et-Oise Gr-3, Prix des Chenes (L) and finished second in the Prix du Gros Chene Gr-3. Bayraan showed speed and was a colt Karim Aga Khan happened to like, so he requested his people inspect Bayraan's younger brother. With their blessing, Karim purchased Blushing Groom privately. They knew Bayraan's little brother was small, standing barely 15 hands high, yet he was put into training with the Aga's other two year olds and Blushing Groom showed surprising speed at his first outing. Runaway Bride's other foals were Chateau Gaillard, a colt by Silver Shark which won 6 races including the Prix du Croix Dauphine Hurdle, Hot Shoes (a filly by Red God) a minor winner in France and Well Done, 5 wins in England and France. Zivato, a colt by Zeddaan, was a disappointment, winning one race at three.

From 6 starts at 2 years, Blushing Groom won 5 of them. At the conclusion of the season he was hailed champion 2YO colt of Europe. In a few months he had escalated from a 2YO Maiden win at Evry to victories in the Prix Morny Gr-1 by 3 lengths defeating Rhino and Water Boy; the Prix Robert Papin Gr-1 by more than 5 lengths from River Dane; the Prix de la Salamandre Gr-1 defeating Assez Cutie and Sunday Guest; and climaxed the juvenile season with 4 lengths victory in the Grand Criterium Gr-1, defeating top rated English juvenile invader J. O. Tobin.

In 1977, Blushing Groom ran four times winning the Prix de Fontainbleau Gr-3 from Water Boy, the French 2,000 Guineas Gr-1 by 3 lengths from Pharly and Hasty Reply, was third in the English Derby Gr-1 to The Minstrel and Hot Grove, tiring over the final furlong, and ended with an unlucky second in the Prix Jacques Le Marois Gr-1 - with a brilliant burst of speed, Blushing Groom got within a half length of Flying Water, but he sure did impress everyone that day.

Blushing Groom's record shows 10 starts for 7 wins, 2 seconds and 1 third. He was voted champion miler in France. His Highness Karim Aga Khan received a substantial offer from John Gaines, owner of Gainesway Farm in Kentucky. After weeks of negotiations a final price was accepted and the son of Red God was shipped to Lexington. It is now history how this small chestnut horse with such character and personality (eventually measuring 15.2 hands) became one of the most influential sires in the world. Here is a list of Blushing Groom's major stakes winners:

Nashwan - Europe's horse of the year, champion 3YO colt, 1st English Derby Gr-1, 2,000 Guineas Gr-1, King George VI & Queen Elizabeth Stakes Gr-1, Eclipse Stakes Gr-1.
Rainbow Quest - Champion 3YO colt and older male in England, 6 wins including the Prix de l'Arc de Triomphe Gr-1, Coronation Cup Gr-1, Great Voltiguer Stakes Gr-3, 2nd Irish Sweeps Derby Gr-1, Grand Criterium Gr-1, 3rd French Derby Gr-1.
Arazi - Champion 2YO colt of Europe and USA, 9 wins including Breeders' Cup Juvenile Stakes Gr-1, Grand Criterium Gr-1, Prix de la Salamandre Gr-1, Prix Morny Gr-1, Prix Robert Papin Gr-2, Prix du Rond-Point Gr-2, Prix du Bois Gr-3, Prix Omnium (L).
Al Bahathri - Champion 3YO filly in Ireland, 1st Irish 1,000 Guineas Gr-1, Coronation Stakes Gr-2 Lowther Stakes Gr-2, 2nd English 1,000 Guineas Gr-1.
Sky Beauty - Champion older mare USA, 1st Coaching Club American Oaks Gr-1, Acorn Stakes Gr-1, Alabama Stakes Gr-1, Mother Goose Stakes Gr-1, Hempstead Handicap Gr-1.
Runaway Groom - Champion 3YO colt in Canada, 1st Travers Stakes Gr-1, Breeders' Stakes Gr-1 Prince of Wales Stakes (Can) Gr-1.
Snow Bride - 1st English Oaks Gr-1, Musidora Stakes Gr-2, Princess Royal Stakes Gr-3.
Ifrad - 1st Arlington Handicap Gr-1.
Nassipour - 1st Rothmans International Stakes Gr-1, Dixie Handicap Gr-2, Seneca Handicap Gr-3.
Sillery - 1st Prix Jean Prat Gr-1, Prix Dollar Gr-2, 2nd Budweiser International Stakes Gr-1 twice.
Rahy - 1st Bel Air Handicap Gr-2, 2nd Middle Park Stakes Gr-1.
Mt. Livermore - 1st Carter Handicap Gr-2, Fall Higweight Handicap Gr-2, Boojum Hcp Gr-3.
You'd Be Surprised - 1st John A Morris Handicap Gr-1, Top Flight Handicap Gr-1, New York Handicap Gr-2, Gallorette Handicap Gr-3, 2nd Hempstead Handicap Gr-1.
Fire The Groom - Beverly D Stakes Gr-1, Wilshire Handicap Gr-2, Santa Anita Breeders' Cup Handicap Gr-3.
Gold Splash - 1st Prix Marcel Boussac Gr-1, Coronation Stakes Gr-1, 3rd French 1,000 Guineas.
Blushing John - 1st French 2,000 Guineas Gr-1, Hollywood Gold Cup Gr-1.
Rosefinch - 1st Prix Saint-Alary Gr-1.
Blush With Pride - 1st Kentucky Oaks Gr-1, Santa Susana Stakes Gr-1, Ashland Stakes Gr-2.
Rosananti - 1st Italian 1,000 Guineas Gr-1, 2nd Italian Oaks Gr-1.
Vigliotto - 1st O.K. Gold Bowl Gr-1 in South Africa.
Groom Dancer - 1st Prix Lupin Gr-1, Prix de Conde Gr-3, Prix Daphnis Gr-3, Prix de Guiche Gr-3.
Digamist - 1st Heinz '57 Phoenix Stakes Gr-1.
Baillamont - 1st Prix Jean Prat Gr-1, Prix d'Ispahan Gr-1, Prix Ganay Gr-1.
Crystal Glitters - 1st Prix d'Ispahan Gr-1 twice, Prix Eclipse Gr-3, Prix Edmond Blanc Gr-3.
Jalmood - 1st Premio Presidente della Repubblica Gr-1 .
Too Chic - 1st Maskette Stakes Gr-1.
Love the Groom - 1st King Edward VII Stakes Gr-2, Goodwood Gordon Stakes Gr-2.

Here is the pedigree of this marvellous chestnut stallion:

				Pharos
			Nearco	Nogara by Havresac II by *Rabelais*
		Nasrullah		Blenheim by *Blandford*
			Mumtaz Begum	Mumtaz Mahal by *The Tetrarch*
	Red God			Pharamond II ex *Selene*
			Menow	Alcibiades
		Spring Adieu		Bull Dog by *Teddy*
			Boola Brook	Brookdale
BLUSHING GROOM				*Rabelais*
			Rialto	La Grelee
		Wild Risk		*Blandford*
			Wild Violet	Wood Violet
	Runaway Groom			Owen Tudor (dam by *Pharos)*
			Tudor Minstrel	Sansonnet
		Aimee		Umidwar by *Blandford*
			Emali	Eclair by Ethnarch, by *The Tetrarch*

Blushing Groom was linebred 5 x 4 x 5 to the great sire Blandford, via two sons and a daughter. He has two sons of Selene in his pedigree, viz. Pharamond II and Hyperion - the latter being sire of English Derby winner Owen Tudor (Hyperion-Mary Tudor II by *Pharos* -Anna Bolena by *Teddy*). Note that Blushing Groom's zig-zag goes back to Blandford! Tudor Minstrel was inbred to Gondolette, grand-dam of Selene. Because Tudor Minstrel's dam Sansonnet was by English Derby winner Sansovino (Swynford-Gondolette), Blushing Groom carries a triple of Gondolette in his make-up; moreover, since Gondolette was by Loved One (son of Pilgrimage) it complemented Pilgrimage's best daughter Canterbury Pilgrim (dam of Chaucer and Swynford). Please remember Blandford is a son of Swynford whereas both Hyperion and Pharos are products of mares by Swynford's half-brother Chaucer.

Furthermore, to complicate matters, Blushing Groom is linebred to champion The Tetrarch via a daughter (Mumtaz Mahal) and a son (Ethnarch, sire of Eclair). Karim's grand-father (the Aga Khan) had purchased Eclair at a broodmare sale in England and was attracted to her because she carried influence from The Tetrarch. It seems Eclair possessed very impressive Vuillier dosage figures. The old Aga Khan already knew about positive genetic affinity between the stallion Gainsborough with mares by The Tetrarch. Eventually Eclair produced Emali and the Aga Khan would send this filly to Tudor Minstrel (grandson of Hyperion, by Gainsborough) to further improve the dosage figures. The result of this match would be a filly named Aimee.

Karim's grandfather did not have access to Hyperion. This was because Hyperion's owner, Lord Derby, took a set against the Aga for selling his best stallions to the United States of America and decided he would never sell Hyperion nominations to help improve his rival's breeding program. At least, that is what Lord Derby admitted in public. Perhaps the real reason was because Lord Derby's racehorses were often defeated by those of the Aga's on the race course, so it may have been a case of jealousy. Rivalry between the two gentlemen was passive, but underneath, intense.

Blushing Groom's broodmare sire Wild Risk was a good French handicapper with a slight club foot. He became France's champion hurdler. Foaled in 1940, Wild Risk was by Rialto, a talented son of St. Simon's son Rabelais. A stakes winner of only two races on the flat, Wild Risk won 9 of 14 starts over the jumps. He was athletic, about 15.3 hands high, and won the Prix Recloses and the

more valuable Prix Sancy. He was placed in the Prix de l'Esperance, La Coupe de Maisons-Laffitte, Prix Beteux and finished third in the Prix des Marechaux. To everyone's consternation, Wild Risk led the French sires' list twice, i.e. 1961 and 1964, and was regularly in the top ten. His female line traced to Quiver, a member of Brown Bess's family. On Blushing Groom's damline sits Polymelus's full sister Lady Cynosure. Might this be important? I believe so, because Polymelus is the sire of Phalaris, the father of Pharos, an ancestor duplicated in Blushing Groom's pedigree.

Blushing Groom carries genetic impact from Brown Bess, the mare to which great stayer Carbine (father of English Derby winner Spearmint) is inbred. Nearco's grand-dam Catnip is by Spearmint, and we know Nearco sired Red God's father Nasrullah. This doesn't mean a whole lot seeing it is way back in Blushing Groom's pedigree but when one includes genetic influence from Bull Dog (Teddy-Plucky Liege by Spearmint) via Boola Brook, Carbine's impact and that of his inbred female ancestor Brown Bess begins to capture our attention and have meaning.

Blushing Groom was leading sire in England in 1989 and leading broodmare sire in 1995. He was third leading sire in France for 1985. Let's briefly examine his female line.

Black Ray, the fifth dam of Blushing Groom, was bred by Mr. J. B. Joel in England. She produced 9 winners including stakes winners Jacopo, Black Domino and champion sprinter Foray, as well as Eclair (filly by Ethnarch) which is the fourth dam of Blushing Groom. Black Ray was only a minor winner yet traced to Our Lassie, one of the best fillies ever sired by Ayrshire. Jacopo won 4 races including the Princess of Wales's Stakes, Molecomb Stakes, and Windsor Castle Stakes. Foray won the Ascot King's Stand Stakes, Champagne Stakes and was by The Tetrarch's son Tetratema.

Eclair won 7 races including the Falmouth Stakes and Lingfield Autumn Oaks. She became the dam of stakes winners Khaled (sire of champion Swaps), Piping Rock, and Eclat as well as Lady Electra and Emali.

Purchased at auction by the Aga Khan in 1942 (from American horse-breeder Marshall Field who had acquired the gray mare's dam from Mr. Joel), Eclair had previously produced three moderate winners sired by three moderate stallions. Eclair's innate potential was noted by the Aga Khan who perhaps knew the mare's cover to Gainsborough had the potential to produce a stakes runner. This ideal match was the first decent mating for the middle-aged mare.

It was rare for the Aga Khan to buy anything at all during World War II whilst living in Switzerland. In fact, he wanted to sell as many horses as possible. Hence, he purchased Eclair to gain her specific genotype and acquire the foal by triple crown winner Gainsborough. Eclair's pedigree had great appeal because in 1936, Mahmoud had won the English Derby in record time for the Aga Khan. Mahmoud's dam, of course, was bred on the Gainsborough/The Tetrarch "nick". Thus, the colt foal born in 1943 became known as Khaled, and went undefeated at two, winning the Ascot Coventry Stakes and Middle Park Stakes. At three, Khaled made a noise in training but it didn't affect his performances. He was runner-up in the English 2,000 Guineas, won the St. James's Palace Stakes and finished a close third in the Eclipse Stakes. Californian horseman Rex Ellsworth bought the horse and raced Khaled's world record-setting son Swaps. Eclat won 2 races including the Duke of Edinburgh Stakes in 1914 and was by the little Hyperion horse Stardust (grandson of Gainsborough), runner-up in the 2,000 Guineas. Eclat showed class by finishing second in the Irish Derby and Eclipse Stakes.

Eclair's filly by stakes winner Umidwar was named Emali. She won 3 races at three, and was a very attractive filly with good shoulders. Among her triumphs were the Ulster Cambridgeshire Handicap in which she defeated many high-class colts; and she finished second in the more valuable

Cambridgeshire Handicap. It was the union between Tudor Minstrel and Emali which set in motion a high class revival of Eclair's family. What did Owen Tudor's son Tudor Minstrel contribute?

The foal from this match was named Aimee. She was treasured by the Aga Khan because of her special pedigree pattern which he had created with the help of Madame Vuillier. The filly's line-breedings were to superior ancestors including Gondolette, the mare which rival breeder Lord Derby had coveted, and of course made use of Gainsborough (grandsire of Owen Tudor).

Everyone knew Gondolette's family. It was a dynasty which had supplied Lord Derby with Hyperion, Pharamond II, Sickle, Selene, Bosworth, Serenissima and Sansovino. The Aga Khan had acquired Gondolette's strain via Hyperion and Sansovino using the stallion Tudor Minstrel. Thus, he was a happy man. (Incidentally, Timeform publications rate Tudor Minstrel as one of the top three turf performers of the entire 20th Century).

When H. H. Karim Aga Khan took over the thoroughbred empire of his grandfather and father Prince Aly Khan (a first-class horseman) he would not have been made aware of the importance of Aimee to the breeding program. Had Prince Aly Khan lived, I daresay Aimee's line would be more influential. When Karim sold Aimee's daughter Runaway Bride, maybe he was unaware of the potential the mare possessed, or maybe he culled her on the advice of his managers. After all, she had not won a race. Yet within her genetic make-up were recessive genes anxious to be exploited.

WILD RISK (Rialto-Wild Violet by Blandford) sire of the dam of Blushing Groom. Leading sire twice in France, his sons include Le Fabuleux, Worden II and Vimy.

It was an outsider who is credited as the breeder of Blushing Groom. However, it was a very shrewd Karim who purchased the Red God colt back as a youngster. Only for the reason that he liked Bayraan was this purchase ever under consideration. From the sale of Blushing Groom, His Highness Karim Aga Khan used some of the proceeds to purchase broodmares and secure all the valuable Dupre mares privately. No doubt his friend and personal trainer Francois Mathet would have urged the Dupre acquisitions.

Infra Red, full sister to Eclair, won two races including the Princess Elizabeth Stakes and finished fourth in the English 1,000 Guineas. She produced four winners including Red Briar (by Owen Tudor), dam of Pipe of Peace, 5 wins including the Middle Park Stakes, 3rd 2,000 Guineas and Derby and a successful sire in Australia. Another filly Riding Rays (by Nearco) won the Nassau Stakes and finished third in the Cheveley Park Stakes.

Riding Rays is the grand-dam of Dubasoff (11 wins 1st American Derby, Arlington Handicap) and fourth dam of Gold River (7 wins including the Prix de l'Arc de Triomphe Gr-1). Infra Red's non-winning daughter Red Ray, a smallish mare light of bone, produced Virginia Water. Paul Mellon, a wealthy American banker sent Virginia Water to Princequillo and got Milan Mill, dam of Europe's horse of the year Mill Reef. What a great colt he turned out to be for trainer Ian Balding! By America's 2YO champion Never Bend, Mill Reef won 12 races including the English Derby Gr-1, Prix de l'Arc de Triomphe Gr-1, Eclipse Stakes Gr-1, Coronation Cup Gr-1, Dewhurst Stakes Gr-1, Prix Ganay Gr-1 and Gimcrack Stakes Gr-2.

I have often wondered whether it would be wise to cross the lines of Blushing Groom with descendants of Mill Reef. A disadvantage might be size - it would depend on the material used. This cross is worth contemplating because it would link together Eclair and her full sister Infra Red in the same pedigree.

Blushing Groom's strain has a genetic affinity with Northern Dancer and Nijinsky II - to be outlined in the next chapter. Because Tudor Minstrel is out of Fair Trial's half-sister Sansonnet, it would appear that Fair Trial's strain via either Petition or Court Martial might cross well and supply classic speed.

This type of mating makes use of a son and a daughter of Lady Juror, with support from her half-sister Mumtaz Mahal, grand-dam of Nasrullah.

Another fast strain which should work with Blushing Groom's is that of Royal Charger's champion son Turn-To (Royal Charger-Source Sucree by Admiral Drake-Lavendula II by Pharos).

His Highness Karim, the Aga Khan.

FAMOUS GENETIC AFFINITIES CALLED "NICKS"

"A good recipe can be spoiled by not adding the right amount of ingredients."

What is a "nick"? An historical review of thoroughbred breeding will inevitably spotlight genetic affinities between two strains that were able to combine in a favorable way to produce stakes class winners. When this happens, racing people call it a "nick" - meaning that the strains nick together in harmony, causing an upgrading of racing class. The term is used in a general sense, and to some extent is misused. A "nick" is not something you analyse via statistics. It is fallible, like most things in the horse business, and is no more than a genetic affinity which works to produce a better athlete than can be achieved normally by a specific stallion: therefore it is valuable since one can use the knowledge *to buy or breed a better class racehorse.* Like all other methods of selection, the technique of copying a "nick" has its high points (Group or Graded stakes success) and its low points (Maiden or Claiming status) determined of course by probability (chance). If a nick is copied and the best genes go unexpressed, then the result will fall below expectancy levels. Alternately, if the nick is copied and the very best genes are expressed, a Graded stakes winner is probable.

There is no point going into lengthy detail about the many famous nicks of last century: instead, let's examine some of the nicks which can be found in today's six-generation pedigrees. Perhaps the first best example to analyze is the Phalaris/Chaucer nick. Without the advantage of covering mares by Chaucer (St. Simon-Canterbury Pilgrim by Tristan), England's premier sire Phalaris may have been little more than an average sire of winners. PHALARIS (Polymelus-Bromus by Sainfoin) was foaled in 1913 and was a muscular horse, well-topped although slightly back at both knees. He was bred by Lord Derby and won 16 races from 24 starts including the Royal Stakes, St. George Handicap, Snailwell Stakes, Challenge Stakes twice, June Stakes and Beaufort Stakes. Phalaris performed over three seasons and remained sound. He was about 10 pounds inferior to the best milers of his generation and might be described as a good handicapper that was well placed by his trainer. First, let's remind ourselves of the male line descent of Phalaris (the Eclipse branch) and list his best sons, whether they were out of mares by Chaucer or not:

- **Bona Vista**
  - **Cyllene**
    - **Polymelus**
      - **PHALARIS - his best colts**
        - **Pharis II (ex Carissima by Clarissimus)**
        - **Pharos (ex Scapa Flow by Chaucer)**
        - **Fairway (ex Scapa Flow by Chaucer)**
        - **Colorado (ex Canyon by Chaucer)**
        - **Manna (ex Waffles by Buckwheat)**
        - **Caerleon (ex Canyon by Chaucer)**
        - **Fastnet (ex Tatonia by Alcantara II)**
        - **Signal Light (ex Ensoleille by Sunstar)**
        - **Bozzetto (ex Bunworry by Great Sport)**
        - **Pharamond II (ex Selene by Chaucer)**
        - **Sickle (ex Selene by Chaucer)**

PHALARIS - his best fillies

Fair Isle (ex Scapa Flow by Chaucer)
Carpet Slipper (ex Simon's Shoes by Simon Square)
Phaona (ex Destination by Desmond)
Priscilla (ex Lammermuir by Sunstar)
Bridge of Allan (ex Spean Bridge by Spearmint)

Of the colts, six of the twelve were out of daughters of Chaucer. Of the fillies, the standout was classic winner Fair Isle, winner of the English 1,000 Guineas and a full sister to Fairway (1st St. Leger) and Pharos (2nd English Derby). Admittedly an "opportunity factor" prevailed because Lord Derby stood Chaucer and retained some of his daughters. Two on the total list were out of daughters of Sunstar, namely the colt Signal Light (4 wins including the Craven Stakes, Malton Plate) and the filly Priscilla (dam of Hyperides which won the Coronation Cup and 2nd English Derby). The best colt ever sired by Phalaris was perhaps Manna whose pedigree is shown in chapter 9, winner of the English Derby and 2,000 Guineas.

Considering Phalaris covered a multitude of mares by different sires, there seems to be a "skew" in favor of the broodmare sire Chaucer "nicking" with the genotype of Phalaris. The pedigree of this affinity is shown below, using a daughter and a son of the duplicated St. Simon:

			Bona Vista
		Cyllene	Arcadia*
	Polymelus		Hampton (to *Queen Mary)*
		Maid Marian	Quiver
PHALARIS			Springfield
		Sainfoin	Sanda
	Bromus		*St. Simon*
		Cheery	Sunrise
			Galopin
		St. Simon	*St. Angela*
	Chaucer		Tristan* (to *Queen Mary)*
with		C. Pilgrim	Pilgrimage
daughters of			
	various female lines..............		

* Arcadia's dam Distant Shore is by Hermit
* Tristan is by Hermit

The match of Phalaris with daughters of Sunstar (Sundridge-Doris) brings in Sundridge's dam Sierra, sister to Sainfoin, broodmare sire of Phalaris. Thus, this genetic affinity has common ancestors on both sides of a runner's pedigree. The Phalaris/Chaucer cross may be described as a real "nick", and was as effective as the renowned Isonomy/Hermit nick so successful towards the end of the 19th century. Two other "nicks" used by horse-breeders in the 20th century were the Gainsborough/The Tetrarch cross and the Blandford/Desmond cross which both resulted in stakes winners far and above the "norm" for respective sires Gainsborough and Blandford.

In the USA, one of the most successful nicks was the temperamental Fair Play with daughters of triple crown winner Rock Sand. Fair Play proved to be an outstanding sire. This affinity came up with runners like Man O'War (Fair Play-Mahubah by Rock Sand) US horse of the year, champion

2 YO & 3YO colt; multiple stakes winner Chatterton (Fair Play-Chit Chat by Rock Sand) 15 wins and successful sire; multiple stakes winner Dunlin (Fair Play-Dona Roca by Rock Sand); multiple stakes winner Mad Hatter (Fair Play-Madcap by Rock Sand) which won 32 of his 98 starts and retired completely sound; stakes winner Sporting Blood (Fair Play-Felicity by Rock Sand) and many others. This nick provided one of the champions of all time - the great Man O'War whose pedigree now follows:

				Australian
			Spendthrift	Aerolite by Lexington *(Glencoe)*
		Hastings		Tomahawk by King Tom *(Glencoe)*
			Cinderella	Manna (USA)
	Fair Play			Doncaster by *Stockwell* *(Glencoe)*
			Bend Or	Rouge Rose
		Fairy Gold		Galliard by *Galopin*
			Dame Masham	Pauline by *Hermit*
MAN O'WAR				Springfield by *Stockwell* *(Glencoe)*
			Sainfoin	Sanda - dam by *Stockwell* *(Glencoe)*
		Rock Sand		St. Simon by *Galopin*
			Roquebrune	St. Marguerite by *Hermit (+ Stockwell) G*
	Mahubah			Hampton
			Merry Hampton	Doll Tearsheet
		Merry Token		MacGregor
			Mizpah	mare by Underhand

The closest linebreeding comes via the duplication of leading sires Galopin (5 x 5), Hermit (5 x 5), Stockwell (5 x 5 x 5 x 7) and the influential mare Pocahontas (dam of King Tom and Stockwell) which was by Glencoe, an ancestor figuring in Lexington's pedigree.

The most successful genetic affinities in North America include the Nasrullah/Princequillo cross which is an extension of Europe's Nearco/Prince Rose nick; the Blue Larkspur/Teddy nick involving Teddy's sons Sir Gallahad III and Bull Dog: the well known Mr. Prospector/Buckpasser affinity; the fairly recent Northern Dancer/Blushing Groom nick; and the compatibility of Menow/Bull Dog or Menow/Sir Gallahad III which wasn't so much a nick (because of low numbers of representatives) but nevertheless, a successful affinity. The newest nick is Northern Dancer with Halo, a cross which whilst successful in the USA has proven to be phenomenal in Japan during the 1990's. The genetic compatibility of Northern Dancer/Halo should be copied, thus creating inbreeding to elite mare Almahmoud. In Europe, we had the famous Bend Or/Macaroni nick; the Hyperion/Donatello II nick which produced Group 1 stayers; the Blandford/Clarissimus and Blandford/Desmond affinities; and many instances of an affinity between parents carrying Sainfoin and his sister Sierra, of which Nasrullah and Royal Charger are representatives.

Outstanding stakes winners bred on the Mr. Prospector/Buckpasser cross are Woodman, (Mr. Prospector-Playmate by Buckpasser) Champion 2YO colt in Ireland - Miswaki (Mr. Prospector-Hopespringseternal by Buckpasser) 6 wins, 1st Prix de la Salamandre Gr-1, Prix Yacowlef, Charles Hatton Stakes, 2nd Prix Morny Gr-1, Fall Highweight Gr-2, 3rd Dewhurst Stakes Gr-1 - Mining (Mr.Prospector-I Pass by Buckpasser) 6 wins from 7 starts, 1st Vosburgh Stakes Gr-1, Coaltown Stakes - and Seeking the Gold (Mr. Prospector- Con Game by Buckpasser) 8 wins $2.3 million 1st Dwyer Stakes Gr-1, Super Derby Gr-1, Peter Pan Stakes Gr-2, 2nd Breeders' Cup Classic Gr-1, Travers Stakes Gr-1, Wood Memorial Stakes Gr-1. This "nick" is shown on page 178.

TUDOR MINSTREL, English champion 2YO and winner of the 2,000 Guineas. His dam is half-sister to Fair Trial.

CANTERBURY PILGRIM, winner of the English Oaks and dam of Swynford and Chaucer.

			Polynesian
		Native Dancer	Geisha
	Raise a Native		Case Ace by *Teddy*
		Raise You	Lady Glory
MR. PROSPECTOR			Nasrullah
		Nashua	Segula by Johnstown (to*Sir Gallahad III)*
	Gold Digger		Count Fleet
		Sequence	*MISS DOGWOOD by Bull Dog by Teddy*
The "nick"			Menow
		Tom Fool	Gaga by *Bull Dog by Teddy*
	BUCKPASSER		War Admiral
		Busanda	*BUSINESSLIKE (Blue Larkspur/Teddy)*
with daughters of			
	various female lines.......		

Note: (1) Miss Dogwood (Bull Dog *son of Teddy* -Myrtlewood by *Blue Larkspur)* is related to Businesslike (*Blue Larkspur*-La Troienne by *Teddy)*

(2) Polynesian is linebred to Sainfoin and his sister Sierra four times.
Nasrullah is linebred to Sainfoin and Sierra.
Count Fleet carries Sierra.
Menow is by Pharamond II, carries Sainfoin.
War Admiral is by Man O' War, carries Sainfoin.

(3) War Admiral carries Swynford's full brother Harry of Hereford, and Spearmint.
Nasrullah carries Swynford and Spearmint.

Now we shall look at the best stakes winners sired by Woodman, Miswaki and Seeking the Gold to see if there are any common denominators, because if we can find them we might discover how prepotency is "working" with their genotypes in the next generation. Only by researching pedigrees can one understand how the biochemistry upgrades speed.

WOODMAN - Best progeny to date are:

	Broodmare sire:	*Second dam's sire:*	*Inbreeding or Duplications:*
Hector Protector	Riverman	Damascus	Pharamond II/Sickle
& Bosra Sham			Native Dancer/Nasrullah
			Bull Dog/Sir Gallahad III.
Hansel	Dancing Count	Victoria Park	Native Dancer/Nearco
Timber Country	Pretense	Swaps	4 x 3 to Swaps
Gay Gallanta	Nureyev	Sir Gaylord	Native Dancer/Nearco
			Mumtaz Begum
Mujtahid	Mill Reef	Majestic Prince	3 x 4 to Raise a Native
Woods of Windsor	Cyane	Rollicking	Nearco, Mumtaz Begum
Mahogany Hall	Majestic Light	Count Amber	3 x 4 to Raise a Native

Summary: The Nasrullah/Princequillo and Royal Charger/Princequillo nick has an affinity.
Reinforced ancestors: Native Dancer, Nearco, Mumtaz Mahal and Selene.

MISWAKI - Best progeny to date:

	Broodmare sire:	*Second dam's sire:*	*Inbreeding or Duplications:*
Black Tie Affair	**Al Hattab**	**Bold Commander**	**Nearco/Mumtaz Mahal**
Urban Sea	**Lombard**	**Espresso**	
Miswaki Tern	**Arctic Tern**	**Gay Lussac**	**Native Dancer/Nearco Mumtaz Mahal**
Waki River	**Mill Reef**	**Val de Loir**	**4 x 4 Princequillo & Nasrullah/Blandford**
Mister Slippers	**Lyphard**	**Riverman**	**Native Dancer/Nasrullah Mumtaz Mahal/Bull Dog Sir Gallahad III**
Playlist	**Northern Dancer**	**Hail To Reason**	**Native Dancer/Nearco Mumtaz Mahal**
Gervazy	**J.O.Tobin**	**Buckpasser**	**3 x 3 Buckpasser**
Mr. Importance	**Nijinsky II**	**Candy Spots**	**Native Dancer/Nearco Mumtaz Mahal/Bull Dog Sir Gallahad III**
Umatilla	**Nijinsky II**	**Sir Ivor**	**Native Dancer/Nearco Mumtaz Mahal/Bull Dog Sir Gallahad III**

Summary: Nasrullah/Princequillo and Royal Charger/Princequillo nick has affinity. Reinforced ancestors: Native Dancer, Nearco, Mumtaz Mahal, Bull Dog

---

SEEKING THE GOLD - Best progeny to date:

Heavenly Prize	**Nijinsky II**	**Riva Ridge**	**Native Dancer/Nearco Mumtaz Mahal/Bull Dog Sir Gallahad III**
Flanders	**Storm Bird**	**Secretariat**	**Native Dancer/Nasrullah Mumtaz Mahal/Selene**
Spain Lane	**Vice Regent**	**Never Bend**	**Native Dancer/Nasrullah Bull Dog/Sir Gallahad III Mumtaz Mahal**
At The Half	**White Gloves**	**Needles**	**Selene/Blenheim**
Seeking Regina	**Cox's Ridge**	**Arts and Letters**	**Nearco/Mumtaz Mahal Sundridge**
Squire Jones	**Private Account**	**Tim Tam**	**3 x 4 Buckpasser Bull Dog/Sir Gallahad III War Admiral/Selene**

Summary: At this early stage, Seeking The Gold might have some affinity with Bull Dog and Sir Gallahad III. He seems to work with Native Dancer's input as well as Nasrullah and/or Royal Charger. Selene (dam of Hyperion) might be another important test.

The nick between Northern Dancer and Blushing Groom can take the form of any of the following combinations, used as examples:

Northern Dancer	**with**	
Nijinsky II	**with**	
Storm Bird	**with**	
Storm Cat	**with**	
Sky Classic	**with**	**MARES by Blushing Groom**
Shadeed	**with**	
Lyphard	**with**	
Caerleon	**with**	
Nureyev	**with**	
Green Dancer	**with**	

Sons of Blushing Groom:

Mt. Livermore	**with**	
Runaway Groom	**with**	
Rahy	**with**	**MARES by Nijinsky II, Northern Dancer, Storm Bird**
Rainbow Quest	**with**	**Danehill, Lyphard, Northfields, Danzig**
Sillery	**with**	

BEST examples to date of the
Northern Dancer/Blushing Groom "NICK"...........

Lammtarra - Champion English 3YO colt 1st English Derby Gr-1, Prix de l'Arc de Triomphe Gr-1 King George VI & Queen Elizabeth Stakes Gr-1.
Peaks and Valleys - Horse of the year in Canada, Champion 3YO in Canada, 1st Rothmans International Stakes Gr-1, Meadowlands Cup Gr-1.
Quest For Fame - 1st English Derby Gr-1, Hollywood Turf Handicap Gr-1, San Luis Obispo Handicap Gr-2.
Kahyasi - Champion English 3YO colt 1st English Derby Gr-1, Irish Derby Gr-1, Lingfield Derby Trial Stakes Gr-3.
Arazi - Champion European and USA 2YO colt, 1st Breeder's Cup Juvenile Gr-1, Grand Criterium Gr-1, Prix de la Salamandre Gr-1, Prix Morny Gr-1, Prix Robert Papin Gr-1, Prix du Rond-Point Gr-3.
Missed the Storm - 1st Test Stakes Gr-1, Astarita Stakes Gr-2, Personal Ensign Stakes, 2nd Prioress Stakes Gr-2, Adirondack Stakes Gr-2.
Pursuit of Love - Champion 3YO sprinter of Europe, 1st Prix Maurice de Gheest Gr-2, Kiveton Park Stakes Gr-3, Ladbroke European Free Handicap (L), 2nd July Cup Gr-1.
Munaaji - Champion Sprinter Germany, 1st Jacobs Goldeine Peitsche Gr-2, etc.
Shamikh - won 1996 Chesham Stakes (Listed) at Royal Ascot .

The potential of the Northern Dancer/Blushing Groom nick is only in its *infancy.* I predict it will be one of the most dynamic "nicks" for the next two decades. Yet another which comes to mind, bred on the Nijinsky II/Blushing Groom affinity, is Graded stakes winner Green Means Go.

Breeders must study the results of Group and Graded stakes and make a note of any consistencies. A "nick" does not have to be by a certain sire out of a daughter of a broodmare sire - it's not that academic! So long as the two individual strains are reasonably close together in the pedigree, it may be described as a "nick" - a harmonious genetic affinity.

The Nasrullah/Princequillo nick was exceptional because it produced many outstanding Group and Graded stakes winners and three juvenile champions. Not every foal born on this nick would inherit the best genes from its elements - probability factors never are generous to a racehorse breeder. Since Royal Charger is three-quarter brother to Nasrullah, it stands to reason that the Royal Charger/Princequillo nick would work, and it did - e.g. US Champion 2YO colt Sir Gaylord (by a son of Royal Charger from a daughter of Princequillo).

Here are a few examples of the NASRULLAH/PRINCEQUILLO nick:

Secretariat - US Horse of the Year, Champion 2YO & 3YO colt, 16 wins from 21 starts, $1.3 million earnings. 1st Kentucky Derby Gr-1 (record time), Preakness Stakes Gr-1, Belmont Stakes Gr-1 NTR, Man O'War Stakes Gr-1, Gotham Stakes Gr-2, Marlboro Cup Hcp Gr-1 .
Riverman - 5 wins from 6 starts, Champion French miler 1st French 2,000 Guineas Gr-1, Prix d' Ispahan Gr-1, Prix Jean Prat Gr-2, Prix Yacowlef, 2nd Champion Stakes Gr-1.
Mill Reef - European horse of the year, Champion 3YO, English Derby Gr-1, Prix de l'Arc de Triomphe Gr-1, Eclipse Stakes Gr-1, Coronation Cup Gr-1, Dewhurst Stakes Gr-1, King George VI & Queen Elizabeth Stakes Gr-1.
Rose Bower - 6 wins $141,034 1st Matron Stakes, Princess Pat Stakes (grand-dam of Miswaki).
Ponte Vecchio - 9 wins $139,263 1st Maskette Hcp Gr-3 (grand-dam of Boundary).
Bold Lad (USA) - Champion 2YO colt, 14 wins $516,465 1st Champagne Stakes Gr-1 etc.
Successor - Champion 2YO colt, 7 wins $532,254 1st Champagne Stakes Gr-1 etc..
Cellini - 6 wins 1st Dewhurst Stakes Gr-1, Tetrarch Stakes, National Stakes, 3rd Irish 2,000 Gns.
King Pellinore - 11 wins 1st Oak Tree Invitational Stakes, Gallinule Stakes, Blandford Stakes, American Hcp, Carleton F Burke Hcp, 2nd St. Leger, 3rd Irish Derby.
Bold Bidder - 13 wins $478,021 1st Charles H Strub Stakes, Hawthorne Gold Cup, Monmouth Hcp.
Quack - 8 wins $514,400 1st Hollywood Gold Cup, California Derby, Will Rogers Stakes etc.
Revidere - 8 wins $330,019 1st Coaching Club American Oaks, Monmouth Oaks, Ruffian Stakes, Gazelle Handicap, Cotillion Stakes, 2nd Beldame Stakes, 3rd Jockey Club Gold Cup.

Next century I foresee numerous Group/Graded stakes winners with the Nasrullah/Princequillo nick appearing twice or thrice in the pedigree; or at least involved with Sir Gaylord which is bred on a similar cross. Because Sky Classic carries the mix of Northern Dancer and Star Kingdom, two ancestors inheriting genetic influence from the speed nick of Hyperion/Tracery, it could indicate he might have a genetic affinity with mares by Blushing Groom or his sons, especially daughters of Mt. Livermore and Runaway Groom. Sky Classic stands at Pin Oak Stud, Kentucky. He won 15 of 29 starts on dirt and turf for $3,240,398 including the Rothmans International Hcp Gr-1, Canadian Maturity Gr-1, Turf Classic Invitational Hcp Gr-1. In Canada, Sky Classic was Champion 2YO, Champion Older horse, Champion Grass Horse and in the USA, Champion Grass Horse.

SUNDAY SILENCE

The most sensational nick at the present time is champion Sunday Silence matched with mares by Northern Taste or Lyphard. Sunday Silence was US Horse of the Year and Champion 3YO colt winning 9 of 14 starts for $4, 968,554 including the Kentucky Derby Gr-1, Preakness Stakes Gr-1, Breeders' Cup Classic Gr-1, Santa Anita Derby Gr-1, Super Derby Gr-1, Californian Stakes Gr-1, San Felipe Handicap Gr-2, 2nd Belmont Stakes Gr-1, Hollywood Gold Cup Gr-1, Swaps Stakes Gr-1 etc. He is almost black with an athlete's body and was foaled in Kentucky in 1986. Sunday Silence is by Halo (twice leading sire USA) from multiple stakes winner Wishing Well by Understanding ($202,655 stakes earner, son of Promised Land) and was bred by Oak Cliff Thoroughbreds. He stands today at Shadai Farm in Japan. His grand-dam Mountain Flower is inbred to Hyperion. Let's have a look at the Sunday Silence/Northern Taste "nick" and see why it works:

				Turn-To by Royal Charger (to MM)
			Hail To Reason	Nothirdchance
		Halo		Cosmic Bomb by Pharamond II
			Cosmah	*Almahmoud by Mahmoud*
	Sunday Silence			Promised Land ex dam by *Mahmoud*
			Understanding	Pretty Ways
		Wishing Well		Montparnesse
			Mountain Flower	Edelweiss
The "nick"				Nearctic ex *Lady Angela*
			Northern Dancer	Natalma ex *Almahmoud by Mahmoud*
		Northern Taste		Victoria Park
			Lady Victoria	*Lady Angela by Hyperion*
	Daughters of			
		various female lines......		

Note:

(1) Duplicates Almahmoud 4 x 5.
(2) Duplicates Mahmoud 5 x 6 x 6 plus his 3/4 sister Mumtaz Begum, grand-dam of Royal Charger.
(3) Provides for linebreeding to Selene, dam of Pharamond II, Sickle and Hyperion
(4) Northern Taste's dam is inbred 3 x 2 to Lady Angela.
(5) Duplication of Nearco and his sire Pharos.
(6) Duplication of Tracery.

In Argentina, Southern Halo has been a successful sire of Group winners and has been brought to the USA to stand the 1996 covering season at Ashford Stud. This stallion damaged his left fore ankle and was Group placed, 5 wins $344,875. He is inbred 3 x 4 to the elite mare Almahmoud and linebred to three-quarter brothers Royal Charger and Nasrullah. He represents the Northern Dancer/Halo nick and has a nice sire pattern with Plucky Liege appearing via three of her sons* plus linebreeding to Mumtaz Mahal and Selene, sourcing considerable maternal strength:

				ROYAL CHARGER by Nearco
			Turn-To	Source Sucree by Admiral Drake*
		Hail To Reason		Blue Swords by *Blue Larkspur*
			Nothirdchance	Galla Colors by *Sir Gallahad III**
	Halo			Pharamond II (*ex Selene)*
			Cosmic Bomb	Banish Fear by *Blue Larkspur*
		Cosmah		*Mahmoud*
			Almahmoud	*Arbitrator*
SOUTHERN				*Nearco*
HALO (1983)			Nearctic	Lady Angela by Hyperion (*ex Selene)*
		Northern Dancer		*Native Dancer*
			Natalma	*Almahmoud by Mahmoud*
	Northern Sea			Dan Cupid by *Native Dancer*
			Sea Bird II	Sicalade
		Sea Saga		Bold Ruler by *NASRULLAH*
			Shama	Lea Lark by Bull Lea (by *Bull Dog*)*

Gainesway Farm's Unbridled (Fappiano-Gana Facil by Le Fabuleux) is a young stallion with a bright future. From his first crop came W.T.Young's homebred Grindstone (1st Kentucky Derby Gr-1, Louisiana Derby Gr-3), and Unbridled's Song (1st Breeder's Cup Juvenile Stakes Gr-1, Florida Derby Gr-1, Wood Memorial Stakes Gr-1). Standing 16.2 hands high with deep chest, strong hocks and decent shoulder angle, Unbridled was Champion 3YO colt of USA winning 8 of his 24 starts to earn $4,489,475 including the Kentucky Derby Gr-1, Breeders' Cup Classic Gr-1, Florida Derby Gr-1, 2nd Preakness Stakes Gr-1, Super Derby Gr-1, Secretariat Stakes Gr-1.

Unbridled is linebred 4 x 4 to the influential mare Aspidistra, and 4 x 4 to the sprinter-miler Rough'n Tumble. He is totally free of Northern Dancer's strain. His broodmare sire is Le Fabuleux, son of Wild Risk. No doubt Unbridled will become a significant sire of classic winners in the future.

Two young sires I especially like in Kentucky have similar pedigree patterns. Robin des Pins (Nureyev-Rare Mint by Key to the Mint) is linebred to Tracery four times - the stallion possessing two daughters of Pocahontas, plus St. Simon's full sister Angelica. Robin de Pins was Champion sprinter-miler of Germany and multiple Group winner in France. Trained by Francois Boutin at Chantilly, he was brilliant enough to be selected as constant workmate for classic winner Kingmambo. He stands in Kentucky at John Phillips's Darby Dan Farm for $6,000 live foal. The other stallion I like is W. T. Young's Mountain Cat (Storm Cat-Always Mint by Key to the Mint) which stands at Overbrook Farm for $7,500 live foal. He is the richest 2YO earner in North America. These young stallions might surprise everyone when they have runners on the track.

Three Chimneys Farm stand Rahy (Blushing Groom-Glorious Song by Halo) and he is certain to get Group 1 and Grade 1 winners from mares by Nijinsky II or other sons of Northern Dancer, thereby duplicating the elite mare Almahmoud. Rahy is sire of champion 3YO filly Serena's Song.

HAIL TO REASON, sire of Halo and Roberto. He hit the board in 13 of his 18 starts and was Champion 2YO colt in USA. He sired 43 stakes winners.

THE MAKTOUM AND NIARCHOS EMPIRES

"The empires of the future are the empires of the mind."
- Sir Winston Churchill

International breeders of the thoroughbred owe a great debt to the *selective breeding* practices stretching from perhaps as long ago as 4,000 years by the Phoenicians, Assyrians, Persians and Egyptian pharaohs to about 1500 years ago by Bedouin tribesmen, rich merchants and the kings and princes of the Middle East region because without the genetic material which evolved and was made available to British horse-breeders (who imported quality purebred *Arabian stallions and mares),* the hybrid racehorse could not have been developed in its existing form. It would not have been possible. Instead, some different hybrid racehorse would have evolved.

Therefore, when Sheikh Mohammed bin Rashid al-Maktoum, a young sheikh from the United Arab Emirate States decided in 1978 he wanted to become involved in the racing and breeding of thoroughbreds, he did so with zest, passion, and the knowledge that his bedouin ancestors helped retain the purity of the *Arabian* breed over centuries past. This man, Crown Prince of Dubai and minister of defence, convinced his elder brothers to become interested in thoroughbred racing. His conceptual ideas led to the initial running of Dubai's World Cup Stakes run on dirt over 1 1/4 miles in mid-March of 1996, valued at an incredible $4 million. Invitations were sent to owners of most of the best horses in training from various countries and this international event was won by Mr. and Mrs. Allen Paulson's champion colt Cigar (by Palace Music-Solar Slew by Seattle Slew) which defeated another American entrant, Soul of the Matter (by Private Terms), owned by musician and composer Burt Bacharach.

Visitors from all over the globe attended this spectacle and the response was very positive. The generosity of the Maktoum family was noted. Sheikh Mohammed is the third born son of Sheikh Rashid bin Said al-Maktoum. His father became the potent ruler of Dubai and with help from British experts, discovered oil and established lucrative oil wells in the southern desert wastelands. Sheikh Rashid was a military man, with dreams and aspirations for his bedouin people. Fighting numerous wars with the aid of his courageous General Abdulla Hussain Lootah (father of Nasser Abdulla), Sheikh Rashid won a deciding battle, and made safe the oil-rich lands that would finance a "renaissance" to the ancient city of Dubai - a city situated on a picturesque inlet called "The Creek". When His Highness Sheikh Maktoum inherited the kingdom from his late father, he began an extensive building program which is perhaps a "model" for all Arab nations to copy. Dubai is very much a modern city with sporting, cultural and historical attractions and offers most of the things that tourists and holiday makers seek. Set on the Gulf, Dubai is a fisherman's haven where luxury yachts can be rented. Major indoor and outdoor sporting events are held annually, including an international golf classic. It is a city of wondrous mosques.

His Highness Sheikh Maktoum is a strong ruler of Dubai. He is vice-president of the United Arab Emirates and although a shy, intelligent gentleman, he is very much admired and loved by his people. Educated in England and responsible for part-westernization of his country, Sheikh Maktoum has invested millions of dollars for education, commerce, communication, shipping and transport and encouraged the employment of foreigners with skills. It is a credit to him and his family that Dubai is today one of the safest cities in which to work or visit in the middle east. It boasts five star hotels, universities, a plethora of international foods and restaurants, plus duty-free

shopping. Thousands of tourists from Europe visit the famous "gold seik" markets. Sheikh Maktoum's government instituted a housing program for his people and I was truly amazed at the government's generosity regarding financial assistance. Sheikh Rashid's sons in order of birth are His Highness Sheikh Maktoum (the ruler of Dubai), Sheikh Hamdan, Sheikh Mohammed and Sheikh Ahmed. They can be proud of the progress made in Dubai in such a short period of time.

Make no mistake, Sheikh Mohammed is a first-class horseman and could get his trainer's licence without any problem if he so desired. He is a self-motivated man with many innovative ideas which he shares with fellow horsemen, be they locals or foreigners. His interests include the sport of *camel* racing, pure-bred *Arabian* racing and a passion for *thoroughbred* racing. Towards the end of 1979, Sheikh Mohammed became seriously interested in purchasing yearlings at Tattersalls and Keeneland when the horse market was already on an upswing, pushed by wealthy buyers including the flamboyant trio of Robert Sangster, Vincent O'Brien and his son-in-law John Magnier. It was at a time when Coolmore Stud was expanding its stallion roster by acquiring many stallion prospects, either Group 1 winning horses or expensive well-bred yearlings with superb damlines. I remember the beginnings of the Maktoum's thoroughbred empire, watching Anthony Stroud and his team search for quality yearling colts and fillies on behalf of Sheikh Mohammed. Next we noticed Michael Goodbody appear with his team acting for H. H. Sheikh Maktoum, as well as Angus Gold and Hubie de Burgh acting for Sheikh Hamdan. There was competition between the Maktoum brothers for the same yearlings, so in the final analysis, it was agreed the brothers would take turns to buy at major sales in order to avoid bidding against one another for the same yearling.

Whenever there was a clash between the Sangster/O'Brien camp and one of the Maktoum brothers for the same yearlings, inevitably for really nice colts, it would make headlines not only in bloodstock magazines or national Daily Racing Form, but also international newspapers and television peak-time news. During the early 1980's, well-bred yearlings fetched millions of dollars and the public packed the July sales venue at Keeneland, hoping for a glimpse of the Maktoums, Robert Sangster and local players wanting action. The spending spree or investment in racehorses by the Maktoum brothers together with certain wealthy Europeans and North Americans, assured the yearling "boom" would last for about seven years.

Stallion fees shot up to dizzy new heights, horse-breeders borrowed money from banks to invest in broodmares and expensive stallion shares, so that the whole frenzied cycle began to feed itself to a point where reality became "fantasy". One yearling fetched a world record price, slightly over a $13 million price-tag: and many yearlings fetched more than $2 million each. Bank managers suddenly believed the horse business was exceedingly legitimate, based on prices paid for yearlings and broodmares in the first half of the 1980's. Who could blame them? Stories of huge profits from consignors penetrated the nooks and crannies of most small towns in or adjacent to the blue grass region of Kentucky - and the word spread interstate, reaching stockbrokers on Wall Street.

But it wasn't to last! Towards the end of the decade, all yearling markets had plummeted downwards, unfortunately to a less than realistic level, and it would send alarm signals to everyone in the bloodstock world and cause bankers to sweat when the final crunch came and broodmare prices decreased.

Stallion share and season prices, temporarily exagerrated over a five year period, came tumbling down and this sector of the market was hit with a resounding thud! However, in the aftermath of the boom and bust, it became clear that thoroughbred investment was indeed, a *legitimate business*, especially when the market readjusted itself, prices picked up and a really solid market was made possible by 1994. The bottom line of this volatile decade for the pragmatic horse investor was to play it "cool" during the height of any future boom. One must learn from past experience.

Among the main reasons why prices for yearlings suddenly declined in the late 1980's were: (1) the dramatic decrease spent by the Maktoum brothers and their friends who by 1987 had collected a substantial quantity of fillies and mares from which they were producing homebred yearlings to race in England; (2) Robert Sangster and Vincent O'Brien temporarily decided to lessen their spending power for expensive yearlings until prospects of either syndicating or selling a classic winning colt were optimistic thereby providing funds for future investment of yearlings; (3) the impending "slow down" of major economies in the western world which could soon affect even Japan's economy; and (4), probably the most important of all, North American and European buyers felt the "boom" market of the 1980's was out of all proportion re "value for money" and they decided to bypass major yearling sales, or at least be quite prudent in their spending power. During the global economic recession of the early 1990's, those fortunate enough to possess ready cash were able to buy bargain-priced yearling fillies and broodmares (probably "steal" might be a better way of putting it) especially during 1992, which seemed to be the bottom of the recession.

Most active at thoroughbred sales were wealthy Japanese buyers from Hokkaido, such as the Yoshida brothers and Dr. Hayata, but there were also Korean investors. Probably the involvement of Korean buyers, new to the marketplace, went unnoticed for a couple of years. Only now do we know they are quite serious about establishing thoroughbred nurseries to produce Korean-bred yearlings. Thus, many outstanding stakes winning fillies and stakes producers departed North American shores for new homes in Asian countries, especially Japan, and to a lesser extent Europe. On a much smaller scale, Australian and New Zealand buyers sneaked some bargains.

One usually makes one's profit when buying in a "bear" market and Asian agents knew they could buy well at major horse auctions and resell at inflated prices within a six months period. For example, the JRA regulations preclude imported foals or yearlings from being entered in Japanese classics, but mares purchased in-foal from abroad may have their Japanese-born foals entered in Japanese classics. These races are worth many milions of dollars, far in excess of purses distributed in North America. Asian agents saw this as an opportunity to buy mares in foal to proven English or American sires: thus, the horse trade began to improve and today we find a healthy, realistic market, again making gradual progress. Immediately prior to the recession, John Gaines sold Gainesway Farm to South African entrepreneur Grahame Beck, while Ben Walden Jr. established Vinery. Farms which were placed on the market slowly began to find new owners.

His Highness Sheikh Maktoum and Sheikh Hamdan (Dubai's finance minister) spent millions developing two farms in Kentucky. Allen Kershaw was appointed general manager of Sheikh Maktoum's magnificent Gainsborough Farm near Versailles (home for Group 1 winner Shadeed). Rick Nicholls was appointed general manager of Sheikh Hamdan's Shadwell Stud, home of English champion 3YO colt Dayjur. The Maktoum brothers' expenditure required to establish and maintain these two large thoroughbred nurseries created considerable employment for Kentucky's trades people and suppliers. Both farms are world-class, state of the art facilities. At time of writing, Sheikh Maktoum's Gainsborough Farm stands four well-bred stallions namely Group or Graded stakes winners Shadeed, Hansel, Sheikh Albadou and Quiet American.

In England and Ireland, the Maktoum brothers established a number of thoroughbred stud farms and created hundreds of jobs for local workers and suppliers. First, Sheikh Mohammed established Dalham Hall Stud at Newmarket, appointing Robert Acton as general manager of the breeding operation and Anthony Stroud as general manager for the racing operation. By 1992, or perhaps even earlier, the Crown Prince of Dubai had a separate racing program in mind, to be operational in Dubai itself. Young colts and fillies would be educated and trained by skilled horsemen from air-conditioned stables specially built to accomodate a racing team in Dubai. The best horse-feed money could buy was imported.

Her Majesty Queen Elizabeth II and His Highness Sheikh Maktoum bin Rashid al-Maktoum photographed at Royal Ascot - photo courtesy of Gerry Cranham. The Queen bred and raced co-champion English 2YO filly Height of Fashion, dam of champion Nashwan.

Sheikh Mohammed decided he wanted to try an experiment by having his better performed two year olds conditioned in the warm climate of the Persian Gulf rather than let them winter in England, Ireland or France. To this end he convinced his eldest brother Sheikh Maktoum to be involved in the "Godolphin" enterprise. Their program was watched with inevitable curiosity by European trainers, and bloodstock agents.

The immediate success of three year old filly Balanchine to win English and Irish classics was a definite boost for Sheikh Mohammed's confidence. Balanchine was exceptional on turf and earned Horse of the Year honors. Other Group winners trained in Dubai would confirm the experiment was a huge success, but naturally enough, it worried some European trainers.

Lammtarra, a Maktoum homebred trained successfully at 2 years by the late Alex Scott who expressed a high opinion about the Nijinsky II colt to Godolphin racing manager Simon Crisford, was sent to Dubai and trained for his assault on the English classics. Lammtarra was not ready for the 2,000 Guineas but he was working like a Group horse in gallops by mid May. This handsome, athletic 16 hands high chestnut son of outstanding race filly Snow Bride, would start in the English Derby and win it on his three year old debut, streaking home for an impressive victory over Tamure and Presenting.

Next, Lammtarra won the King George VI & Queen Elizabeth Stakes Gr-1 at Royal Ascot from the smart colt Pentire, then followed it up in early October with a narrow victory on a rain-affected track to win Longchamp's valuable Prix de l'Arc de Triomphe Gr-1 from a star-studded field of Group winners. Runner-up in the 1995 Arc was Freedom Cry, with impeccably-bred Swain a gallant third. Sheikh Mohammed must have been exceedingly proud of Lammtarra's victories. The colt was shown in the racebook as being owned by Sheikh Mohammed's nephew, Sheikh Maktoum's eldest son. Nevertheless, it was a family triumph to be enjoyed, as is always the case when victory comes in a race like the Derby. In 1996, Godolphin's stayer Classic Cliche won the Ascot Gold Cup.

Collectively, the Maktoum brothers have dominated racing in Europe since the mid-1980's starting with the little brown colt Shareef Dancer, winner of the Irish Sweeps Derby Gr-1 and the class race mare Awaasif; but it has come at great cost. They have won numerous classic races with homebreds as well as sales horses and have been very lucky to win the English Oaks Gr-1 several times, Moonshell being their latest, and the Diesis filly Pricket being runner-up in 1996 to the Caerleon filly Lady Carla. Among the most popular horses they have raced are Dayjur, Cadeaux Genereux, Wolfhound, Arazi, Awaasif, Al Bahathri, Indian Skimmer, Diminuendo, Rahy, Intrepedity, Swain, Oh So Sharp, Sonic Lady, Nashwan, Unfuwain, Ezzoud, Green Desert, Shadeed, Zilzal, and Balanchine. The list is too extensive to mention all the good ones.

I suppose Nashwan (Blushing Groom-Height of Fashion by Bustino) would be the best horse raced by the Maktoums to date. Europe's Horse of the Year and champion 3YO colt, Nashwan had acceleration and was one of the better Derby winners since 1950. He stands at Shadwell Stud, north of Newmarket. Probably the best two year old the Maktoums have raced would be Arazi, another Blushing Groom colt with inbreeding to Wild Risk. Originally owned by Allen Paulson, and selected as a weanling by his manager Ted Carr, Arazi won a string of Group races at two in France before the Maktoums bought a half share in the undefeated colt a few weeks prior to his sensational win in the Breeders' Cup Juvenile Stakes Gr-1. Retired to Dalham Hall Stud, Newmarket, after winning two stakes at three, Arazi has his first crop to race soon. I recently inspected some of his first and second crop youngsters and will admit he is likely to make an influential sire. I liked them a lot.

Arazi's pedigree reveals a mixture of speed and stamina with an inbreeding to Wild Risk which carries Polymelus's full sister Lady Cynosure (thus, utilizing a brother and sister combination).

				Nearco
			Nasrullah	Mumtaz Begum *by a son of Blandford*
		Red God		*Menow*
			SPRING RUN	Boola Brook by Bull Dog
	Blushing Groom			*Rialto*
			Wild Risk	*Wild Violet by Blandford*
		Runaway Bride		Tudor Minstrel - *g'son of Hyperion*
			Aimee	Emali
ARAZI (1989)				*Nearco*
			Nearctic	Lady Angela by *Hyperion*
		Northern Dancer		*Native Dancer*
			Natalma	Almahmoud
	Danseur Fabuleux			*Wild Risk*
			Le Fabuleux	Anguar
		Fabuleux Jane		Raise a Native *by Native Dancer*
			Native Partner	Dinner Partner by *TOM FOOL by Menow*

His pedigree shows 3 x 4 to a daughter and son of influential French stallion Wild Risk; the genetically related ancestors Spring Run (Menow-Boola Brook by Bull Dog) and Tom Fool (Menow-Gaga by Bull Dog); linebreeding to Selene whose sons Pharamond II, Sickle and Hyperion appear further back in the pattern; linebreeding to a son and daughter of Pharos (involving duplication of Nearco); and reinforcement of Blandford and Mumtaz Mahal.

I remember Arazi as a weanling at Keeneland when he sold for more than $300,000. He was an athletic colt, mind you not exactly perfect in his forelegs, but glowing with "presence" and when he walked away or towards you, he displayed an over-stride. Although he resembled his sire in some ways, he also inherits traits from his broodmare sire Northern Dancer. I inspected Arazi again on a recent visit to Dalham Hall Stud, Newmarket. He has matured into a very nice stallion. From 14 starts Arazi scored 9 times for career earnings of $1,300,206. Except for bad luck at three, he might have doubled this amount.

His dam Danseur Fabuleux was stakes-placed in France, 2nd Prix de Minerve Gr-3 and produced stakes-placed winner Fortroe (by Forty Niner), 2nd Prix des Lilas. The next dam is stakes winner Fabuleux Jane, 4 wins in France including the Prix de Pomone Gr-3, Prix de la Porte de Passy, Prix des Belles Filles, 2nd La Coupe de Maisons-Laffite Gr-3, 3rd French Oaks Gr-1, Prix Vermeille Gr-1 and Prix de Royallieu Gr-3. She produced stakes winners Doubles Partner and Fabuleux Dancer and is a half-sister to European 3YO champion sprinter Ajdal and Flying Partner - 4 wins $341,392 which won the Fantasy Stakes Gr-1, Santa Ynez Stakes Gr-2, and finished third in the C.C.A. Oaks Gr-1, and Kentucky Oaks Gr-1.

Sheikh Maktoum owns Woodpark Stud, Ireland, where Kate Ryan and John Brady raise many well-bred foals and yearlings. These are extremely nice people and love what they are doing. At Gainsborough Stud, Newbury, Michael Goodbody (general manager of Sheikh Maktoum's thoroughbred empire) is assisted by stud manager Paddy Bowles, an astute horseman and as Irish as they come. Gainsborough Stud, as the name implies was once the home of a triple crown winner and is situated at Woolton Hill, Newbury. Joe Mercer is Sheikh Maktoum's racing manager. The thoroughbred breeding operations of the four Maktoum brothers comprise in excess of some 600 mares. They are the architects of a racing and breeding empire on a much bigger scale than any of the great breeders of the past. Sheikh Mohammed was leading owner in England for five straight years (1985-1989) and for another three years (1991-1993). He raced Breeders' Cup winners Pebbles and In the Wings. The Maktoums have reached the stage where they must sell annually many well-bred colts and fillies to keep the horse population to a manageable number, inevitably selling future stakes producers at auction, such is the quality of broodmares in their possession. They stand their best performed Group winning colts either in Europe or North America, thus having access to sire material of their own making.

THE NIARCHOS EMPIRE

When Marcel Boussac's empire folded, French Government officials ordered his real estate to be liquidated, including his thoroughbred farms. The magnificent, historic Haras Fresnay-Le-Buffard was purchased in November of 1979 by Greek shipping tycoon Stavros Niarchos. Situated near the township of Falaise in the Normandy region of France, this famous stud farm was overtaken by German troops during World War II. Marcel Boussac was leading owner-breeder in France for many years, racing champions like Tourbillon, Pharis II, Corrida, Coronation, Djebel. His classic winners include Galcador, Asterus, Adargatis, Asmena: other talented stakes winners include Astronomie, Apollonia, Zariba, Caravelle, Abdos and Akarad. Valuable broodmares roamed Fresnay-Le-Buffard's lush green paddocks including descendants of the truly great mare Frizette.

Stavros Niarchos's entry into the thoroughbred world is remembered as beginning in 1953 when he raced a few horses in England. His initial big stakes winner was Dactylographer, son of champion Secretariat, which won the Sandwich Stakes and Observer Gold Cup. From Keeneland sales he purchased Gay Missile, dam of stakes winner Gay Mecene, and asked Peter Walwyn and Sir Philip Payne-Gallwey to buy yearlings there again in 1978. They came away with a Northern Dancer yearling colt from Special by Forli named Nureyev, after giving $1,300,000 for the colt which set a Keeneland record for a yearling. In France, Nureyev won his only start at 2 in the Prix Thomas Bryon Gr-3 over 1500 metres and was put away to develop. At his three year old debut he easily won the Prix Djebel over 1400 metres like a very classy colt and was sent to Newmarket to run in the 2,000 Guineas Gr-1. It was with mixed feelings that Stavros Niarchos saw his expensive colt first over the line in the mile classic, only to lose the race on protest to Known Fact (second past the post) on an objection from the jockey of third-placed Posse. The protest was for alleged interference over the final furlong and Nureyev was disqualified from first to last. The colt never ran again and was syndicated to stand in Kentucky at Walmac International where he would sire 91 stakes winners with progeny earnings of $31 million. Fortunately, Stavros Niarchos retained shares in the stallion for his own breeding program.

More stakes winners followed and stallions began to stand at Fresnay-Le-Buffard once again. Meantime, Niarchos decided to purchase a farm in Kentucky to board some of his stakes-winning fillies, a property adjacent to Lane's End Farm. He gradually began to form a quality racing team and had them trained by Francois Boutin at Chantilly. Sir Philip Payne-Gallwey of the BBA was appointed racing manager, Tim Richardson was appointed stud manager and Niarchos's attractive daughter Maria was held responsible for cost controls, marketing and promoting the stallions, and directing the farm's breeding program.

In the 1980's, some marvellous broodmares were acquired and outstanding homebreds won Group races. Sir Philip Payne-Gallwey selected a bay filly at Keeneland's July sales which would result in none other than Coup de Folie (Halo-Raise the Standard by Hoist the Flag out of Natalma). An expensive yearling, she was consigned by Windfields Farm and was out of a half-sister to Kentucky Derby winner Northern Dancer. I saw this filly sold at Keeneland and congratulated Philip many months later on the purchase. The yearling market by this time was extremely competitive - way above expectations of both buyers and sellers.

Coup de Folie won 4 races at 2 and 3 years including the Prix d'Aumale Gr-3, Prix de Saint-Cyr (Listed), and finished third in the Prix Marcel Boussac Gr-1. Sent to the USA she finished third in the Royal Heroine Stakes before retirement. I asked Francois Boutin what he thought of Coup de Folie and he told me he thought the filly was perhaps unlucky not to have won the Prix Marcel Boussac. He had a high opinion of this Halo filly which was inbred 3 x 3 to Almahmoud.

For Stavros Niarchos, Coup de Folie proved a goldmine. Maria decided to match her with America's proven sire Mr. Prospector, based at Claiborne Farm. The first foal from this union produced French champion 2YO colt Machiavellian, 4 wins from 7 starts earning $357,023. He won the Prix Yacowlef (L) defeating Jade Robbery and Septieme Ciel, the Prix Morny Gr-1 defeating Qirmazi, Mill Lady and Ozone Friendly and gave a dazzling display to take the valuable Prix de la Salamandre Gr-1 from Qirmazi, Ernani and Rock City thereby retaining his unbeaten juvenile status and earning him 2YO championship honors. At three, Machiavellian started four times. He won the Prix Djebel (Listed) over 1400 metres and was dispatched to Newmarket. In the English 2,000 Guineas Gr-1 he ran as one of the favorites and finished on strongly, failing to catch the winner Tirol, with Anshan third followed by Rock City and Septieme Ciel. Sold to the Maktoum brothers for a considerable amount of money, Machiavellian resides at Dalham Hall Stud.

The second union with Mr. Prospector produced Machiavellian's sister Coup de Genie, champion 2YO filly of France. She won the Prix Morny Gr-1, Prix de la Salamandre Gr-1 and would finish third in the English 1,000 Guineas Gr-1. Coup de Folie was matched with French champion Irish River and produced Exit To Nowhere, now resident stallion at Fresnay-Le-Buffard. This tall horse won 4 races including the Prix Jacques Le Marois Gr-1 (1600m), Prix Thomas Bryon Gr-3, Prix Edmond Blanc Gr-3 and was placed in the Prix d'Ispahan Gr-1. He is a colt linebred to Source Sucree, dam of champion Turn-To. Coup de Folie then produced Hydro Calido, a filly by Nureyev which won the Prix d'Astarte Gr-2, Prix d'Angerville (L), and was 2nd French 1,000 Guineas Gr-1, and Prix de Sandringham Gr-3.

Stakes winner Whakalyric (by Miswaki-Lyrism by Lyphard from Pass a Glance by Buckpasser) is another high-class producer. She is inbred 3 x 3 to champion Buckpasser and won the Prix du Calvados Gr-3 and stakes-placed in both the Prix de la Salamandre Gr-1, Prix de la Foret Gr-1 before producing French Derby Gr-1 winner Hernando, a colt by Niniski. This mating which duplicated Northern Dancer 3 x 4 via two sons also includes linebreeding to a son and daughter of Pharos, a son and daughter of Native Dancer, plus three-parts siblings Royal Charger and Nasrullah. Other attractive foals from Whakilyric are Prix Lupin G-1 winner Johann Quatz (by Sadler's Wells) a horse with lots of fast twitch muscle fiber, winner in the USA of the Colonel F W Koester Handicap Gr-2, Fiddle Isle Handicap (L), 2nd Breeders' Cup Mile Gr-1 and 3rd Arlington Million Gr-1; and Walter Willy, brother to Johann Quatz.

Rare Mint (Key to the Mint-Another Treat by Cornish Prince) produced five stakes winners for Stavros Niarchos and Maria. They include Robin des Bois (by Nureyev), winner of the Criterium d'Evry and leading freshman sire in Argentina where in his first crop he sired the winner of the 1995 Argentine 2,000 Guineas Gr-1 and Derby Gr-1; Robin des Pins (by Nureyev) Champion Sprinter of Germany, 6 wins 1st Jacobs Goldene Peitsche Gr-2, Prix du Palais Royal Gr-3, Prix de Saint-Georges Gr-3, 2nd Newmarket Challenge Stakes Gr-2 and 2nd Hollywood Turf Express Handicap (L) which stands at Darby Dan Farm (Fresnay retain shares in this sprinter); and the filly Mystery Rays (by Nijinsky II) which won the Prix Fille de l'Air Gr-3 and Prix Minerve Gr-3.

Miesque, a filly by Nureyev from Pasadoble by Graustark's son Prove Out, was the brightest star created from Maria's program. Foaled in 1984, this strong bay filly was sent to Francois Boutin at Chantilly and the reports about her were positive. From three seasons of campaigning against colts

Miesque accumulated more than $2 million from 12 wins in Europe and the USA. She twice won the Breeders Cup Mile Gr-1 against colts. Miesque was voted Champion 2YO and 3YO filly in Europe, Champion older mare, Champion Miler in France and the USA twice. She won the French 1,000 Guineas Gr-1, Prix Marcel Boussac Gr-1, Prix de la Salamandre Gr-1, Prix Jacques Le Marois Gr-1 twice, Prix du Moulin de Longchamp Gr-1, Prix d'Ispahan Gr-1, Prix Imprudence Gr-(L), was 2nd French Oaks Gr-1, Prix du Moulin Gr-1; won the English 1,000 Guineas Gr-1 and was runner-up in the Queen Elizabeth Stakes Gr-1. Her performances at Group 1 level made her one of the great fillies of the century and she has already produced Kingmambo (c. by Mr. Prospector) and East of the Moon (f. by Private Account). Kingmambo stands at Lane's End Farm, Kentucky. From 5 wins he earned $734,804 with victories in the French 2,000 Guineas Gr-1, Prix du Moulin de Longchamp Gr-1, St. James's Palace Stakes Gr-1 and was runner-up to Zafonic in the Prix de la Salamandre Gr-1. He finished third to Bigstone and champion Barathea in a hotly contested Queen Elizabeth II Stakes Gr-1 with champion filly Sayyedati behind. East of the Moon was Champion 3YO filly of France. She won 4 races including the French 1,000 Guineas Gr-1, French Oaks Gr-1, Prix Jacques Le Marois Gr-1 and was runner-up in the Prix du Moulin de Longchamp Gr-1.

In 1995, Francois Boutin lost the battle against cancer. A great racehorse trainer, married to William T. Young's daughter Lucy, he was a tremendous asset to the Niarchos racing empire and was sorely missed. Then in May 1996, Stavros Niarchos passed away, leaving control and the direction of Fresnay-Le-Buffard to his daughter Maria, a keen judge of racehorses and a lady who understands international pedigrees. Maria likes to consider all possibilities before deciding on a specific mating for the mares and I have no doubt she will not only maintain the success rate achieved by her father, but might improve upon it. The Niarchos family's racing manager is now Alan Cooper, son of Tom Cooper. He is best described as nature's gentleman. In the USA, Kingmambo looks set for a stunning career as a sire at Lane's End Farm. Stallions which are sons of great mares usually sire Group 1 winners. Miesque's female line traces to Neriad, dam of 13 foals including French Champion 3YO filly Comtesse de Loir, winner of the Prix Saint-Alary Gr-1, and runner-up in the French Oaks Gr-1, Prix Vermeille Gr-1 and Prix de l'Arc de Triomphe Gr-1. This is the branch which produced stakes winners Heron Cove, Firyal, Silver Supreme ($645,545 1st Booklyn Handicap Gr-1) and Massarat. Miesque's dam happens to be bred on the successful mix of

			Nearco
		Nearctic	*LADY ANGELA* by Hyperion*
	Northern Dancer		Native Dancer
		Natalma	Almahmoud
	Nureyev		Aristophanes by *Hyperion*
		Forli	Trevisa by Advocate (to *Tracery)*
	Special		Nantallah
		Thong	Rough Shod II
MIESQUE (1984)			Ribot (to *Papyrus by Tracery)*
		Graustark	Flower Bowl by *ALIBHAI**
	Prove Out		Bold Venture
		Equal Venture	Equal
	Pasadoble		Fine Top
		Sanctus	Sanelta (to Bull Dog's sister)
	Santa Quilla		Princequillo (dam by *Papyrus)*
		Neriad	Sea Change by Count Fleet

Ribot with Princequillo, exactly the same mix found in the pedigree of US Champion three year old Key to the Mint. It is not surprising then, that Maria and Tim Richardson chose to send Rare Mint (by Key to the Mint) to Nureyev to copy Miesque's pedigree pattern and acquired from this match two fast horses, Robin des Bois (successful sire in Argentina) and the brilliant sprinter Robin des Pins (by Nureyev) which are bred on identical lines to Miesque's pedigree pattern.

The Ribot/Princequillo cross is based on inbreeding to Derby winner Papyrus, via Ribot's grand-dam Barbara Burrini, and via Princequillo's dam Cosquilla - both by Papyrus, best son of Tracery. Sainfoin's grandson Tracery appears again via two genetically related ancestors in Miesque's pedigree i.e. Nearctic's dam LADY ANGELA bred on the Hyperion/Tracery mix just like Graustark's broodmare sire ALIBHAI (Hyperion-Teresina by Tracery). Surely this potent reinforcement of the same genetic mix had something to do with an upgrading for speed. Miesque is not the only Group 1 winner with this mix, but she certainly is the best example.

Forli, champion of Argentina, is linebred to Lady Juror (her dam by Sundridge, son of Sierra) so he provides Sainfoin's full sister in the pattern. Nureyev's grand-dam Thong is by Nantallah, which carries Sir Gallahad III whereas Sanctus traces to Noor Jahan, full sister to Sir Gallahad III. The pedigree also reveals linebreedings further back to English Oaks winner Canterbury Pilgrim and to Hyperion's dam Selene. In my opinion, Miesque has such a strong pedigree pattern that she can start a powerful branch of her family in her own right. She is an elite mare.

The future of Fresnay-Le-Buffard is in the hands of capable professionals and we can expect more commercial stallions to be raised in this peaceful region of Normandy. Stavros Niarchos's daughter saw the beginning of a new era when homebred Spinning World (by Nureyev) won the 1996 Irish 2,000 Guineas Gr-1 after an unlucky second in the French 2,000 Guineas Gr-1. With determination and dedication, Maria is certain to breed Group 1 winners. The glory days of Fresnay-Le-Buffard are now in full swing and Maria's dreams will materialize.

Ribot's son TOM ROLFE, won 16 races for $671,297 including the Preakness Stakes, American Derby, and Arlington Classic. He sired champion Hoist the Flag.

SIGNIFICANT BREEDERS OF THE 20TH CENTURY

"Progress, therefore, is not an accident, but a necessity. It is part of nature." - Herbert Spencer

There are so many heartbreaks, so many set-backs, so many things that unexpectedly can go wrong when you become involved with thoroughbreds that a person needs to be *fully* committed to a program to breed stakes runners. One's initial fascination quickly wears thin when things don't go to plan. A horse-breeder may have to wait ten years before the good one, the big horse, comes into the shedrow or stable. Yet when it happens, it's fantastic! Imagine how proud Allen Paulson must feel knowing he bred a colt as good as Cigar? Suddenly *the whole of the US of A owns his horse - then the whole world owns his horse!* Offers come from everywhere asking, begging Allen Paulson to start Cigar at their racetrack. Fans want to buy photographs. They want to read about "their" horse. The media frantically keep pace with any latest development about Cigar. They phone Billy Mott and Allen Paulson on a daily basis, and sometimes twice a day.

It happened years ago with Man O'War. It happened with Secretariat. It happened only a year or so ago when Holy Bull was pushing up racetrack attendances. Now, it's the mighty Cigar.

Thousands of miles away "down under" the same thing is happening. In Australia during 1995/96 a wonder colt called Octagonal (Zabeel-Eight Carat by Pieces of Eight) demonstrated speed on turf over the classic Derby distance to stake his claim as the best three year old in the world. Octagonal would scare the living daylights out of 1996 English Derby winner Shaamit. In the Rosehill Guineas Gr-1 2000m (1 1/4m) Octagonal ran a 1-59 and change then came out and won the AJC Derby over 2400 metres. He is a colt that captures the heart and spirit of a whole nation with his courage, and is a fantastic drawcard to boost racetrack attendance. He's a champion!

Who breeds our best racehorses? What kind of people are they? Breeders of Group 1 and Grade 1 winners are usually very special people in our industry. They are dedicated horse-lovers who by means of careful selection are able to upgrade families to Group 1 status.

These people accept challenge. Each and every one of them possess some form of competitve spirit. Underlying all this, they seek personal satisfaction which comes with being successful. Their self esteem rises to new heights. For example, take Bill Young at Churchill Downs in May. When William T. Young won the 1996 Kentucky Derby Gr-1 with his homebred colt Grindstone (Unbridled-Buzz My Bell) he was seen to be emotionally overwhelmed by the closeness of the Kentucky Derby finish. His colt flashed home down the center of the stretch and got to the wire with Cavonnier. The photo finish showed Grindstone a very narrow winner and the moment the colt's number went up on the board, W.T. got nervous - real nervous.

Emotion gripped him: his nerves tingled: for a moment, W.T. was lost for words. Deep inside he must have felt exceedingly proud. His family were there to share the excitement, for they knew the odds of anyone breeding or owning the 1996 Kentucky Derby winner were about 37,000/1. Only one foal can win the classic.

W.T.'s next thoughts would have been for the jockey Jerry Bailey, the trainer D.Wayne Lukas, the colt's handler and everyone at Overbrook Farm. Bill Young always thinks of others. He's a special man in our business.

This chapter briefly examines famous horse-breeders of the 20th century. It is like an honor roll. It makes sense to analyse the methods and techniques used by successful racehorse breeders so that perhaps something can be gained from the exercise. Only the truly great breeders are covered in this chapter. We will deal with the main characters, one by one, starting with the first half of the 20th century.

The most significant aspect, common to nearly all the great breeders of the past, is that they possessed at least one outstanding foundation mare. In the instance of His Highness the Aga Khan, he was fortunate enough to have six influential taproot mares that constantly gave him stakes winners. The best foundation mare was Mumtaz Mahal, Champion 2YO filly of her year and purchased at auction from Sledemere Stud, a small breeding establishment with quality females. Lord Derby based his success on three mares that became the foundation of his racing stable. Without them, he would have failed. Italy's famous breeder/owner/trainer Federico Tesio searched all of Europe to acquire foundation mares and was fortunate enough to acquire four outstanding foundation mares, one of which was Catnip, grand-dam of champions Nearco and Niccolo dell'Arca.

A genealogist like myself truly appreciates the efforts of professional breeders who attained their goals by breeding a champion racehorse. One has to admire their skills and their dedication. Here now, is the list of the significant breeders for the period 1900-1950:

ENGLAND and IRELAND:

LORD DERBY (the Earl of Derby) - breeder of Hyperion, Pharos, Fairway, Heliopolis, Gulf Stream, Sickle, Selene, Pharamond II, Hunter's Moon, Alycidon, Tranquil, Sansovino, Bosworth, Plassy, Chaucer, Swynford, Ferry, Stedfast, Keysoe, Colorado, Fair Copy, Sun Stream, Garden Path, Watling Street, Tobogga, Swallow Tail, Canyon, Bobsleigh, Herringbone, Tide-way, Keysoe, Ferry, and Fair Isle. With the help of studmaster and pedigree consultant Walter Alston, Lord Derby developed many powerful female lines into classic winning lines. From a study of his techniques, his main success can be traced to the purchase of three outstanding mares that literally rocketed him to stardom as a breeder.

Those three mares were Canterbury Pilgrim, winner of the English Oaks (dam of Swynford and Chaucer); Gondolette, a foundation mare whose line kept producing stakes winners every generation, supplying English Derby winners Sansovino and Hyperion; and Anchora, which when matched with Chaucer produced Scapa Flow, dam of successful sires Pharos and Fairway.

Lord Derby duplicated influential mares in most of his matings. His clever studmaster Walter Alston encouraged him to test matings reinforcing Pilgrimage (2,000 Guineas, 1,000 Guineas winner) via her colt Loved One and filly Canterbury Pilgrim; matings reinforcing full brothers Fairway and Pharos - purposely doubling up on their dam Scapa Flow; and matings reinforcing dominant stallions like St. Simon and Stockwell. Few farms have as good a record of producing classic winners as Lord Derby's Stanley House Stud (Woodland Stud) and Plantation Stud.

Lord Derby's name will be remembered for a long time because he bred two of the best stallions in the early part of the 20th century namely Pharos and Hyperion. Without Pharos or Hyperion we wouldn't have Nearco, Northern Dancer, or Nasrullah. Lord Derby was a leading owner-breeder for many years in England. In addition, he was a successful owner in France.

Here are two typical matings designed by Walter Alston for Lord Derby's breeding program. He attempted to exploit the best genes as follows:

			Vedette
		Galopin	*Flying Duchess*
	St. Simon		King Tom *ex Pocahontas*
		St. Angela	Adeline
	Chaucer			Hermit*(brother to Chanoinesse)
	Canterbury Pilgrim	Tristan	Thrift
			The Palmer (brother to Rosicrucian)
		Pilgrimage	*Lady Audley*
SELENE			Bona Vista
		Cyllene	Arcadia (to Hermit*, br. to Chanoinesse)
	Minoru		Friar's Balsam - Hermit*
		Mother Siegel	mare Galopin (to Chanoinesse*)
Serenissima			See Saw
		Loved One	*Pilgrimage by The Palmer*
	Gondolette		Doncaster by Stockwell ex *Pocahontas*
		Dongola	Douranee by *Rosicrucian*

Selene earned the equivalent of $70,000 at the beginning of this century - a lot of money in those days. She was tough and genuine and produced Sickle, Pharamond II, Hyperion, Moon Priestess, All Moonshine, New Moon and Hunter's Moon. She is the ancestress of New Zealand's leading stallion Sir Tristram, which sired a record number of Group 1 winners in the southern hemisphere. Selene is inbred 3 x 4 to classic winner Pilgrimage; linebred 4 x 5 x 5 to full brothers Rosicrucian and The Palmer; linebred to Derby winner Hermit three times, plus his full sister Chanoinesse; and carries linebreeding to Pocahontas via King Tom and Stockwell. The pedigree reinforces her best ancestors.

			Bay Ronald
		Bayardo	Galicia
	Gainsborough		St. Frusquin by *St.Simon*
		Rosedrop	Rosaline
Hyperion			*St. Simon*
		Chaucer	*Canterbury Pilgrim ex Pilgrimage*
	Selene		Minoru *by Cyllene*
		Serenissima	Gondolette by Loved One *ex Pilgrimage*
GULF STREAM			Polymelus *by Cyllene*
		Phalaris	Bromus
	Fairway		*Chaucer (St.Simon-Canterbury Pilgrim)*
		Scapa Flow	Anchora
Tide-way			John O'Gaunt
		Swynford	*Canterbury Pilgrim ex Pilgrimage*
	Drift		Neil Gow
		Santa Cruz	Santa Brigida *by St. Simon*

Gulf Stream won 5 races including the Gimcrack Stakes, Eclipse Stakes, Craven Stakes, and was runner-up in the English Derby. He was sold to South America where he became leading sire of winners. His pedigree shows heavy reinforcement of Pilgrimage via her daughter Canterbury Pilgrim and via her son Loved One. Walter Alston has used 3 x 4 inbreeding to Chaucer, and added Chaucer's half-brother Swynford. The pedigree tells us Gulf Stream might make a prepotent sire.

H. H. THE AGA KHAN and his son PRINCE ALY KHAN - they bred Mahmoud, Nasrullah, Rivaz, Malindi, Petite Etoile, Tulyar, Palestine, Noor, Saint Crespin III, Blenheim, Bahram, Dastur, Umidwar, Hindostan, Mumtaz Begum, Mah Mahal, Nizami, Nilo, Diableretta, Udaipur, Yasmin, Venture VII, Rustom Pasha, Masaka, Gallant Man, Mirza II, Badruddin, Ut Majeur, Amante, Alishah, Theft, Tehran, Aimee, Nathoo, Taj Akbar, Dust Devil, Khaled, Costaki Pasha, Ujijji, Taboun, Turkhan, Butiara, Taj Mah, Fiorentina, Rose Royale II, Claro and Firdaussi.

His Highness the Aga Khan ventured into the business with the help of William Hall Walker (Lord Wavertree) and George Lambton, the trainer of Lord Derby's horses. He was mystified by Colonel Vuillier's "dosage" formula, based on collecting genetic influence from superior stallions and a mare named Pocahontas. The Aga Khan acquired outstanding fillies at the yearling sales, raced them, and then used the Vuillier dosage method to select suitable stallions. He bred Nasrullah, leading sire on two continents, and established six successful female lines, namely Mumtaz Mahal, Uganda, Eclair, Qurrat-al-Ain, Friar's Daughter and Mirawala. The Aga's son Prince Aly Khan was a first-rate horseman and a very good judge of yearlings (his close friend Alec Head confirmed this information for me) but tragically Prince Aly Khan was killed in a motor car accident whilst in the prime of his life.

The Aga Khan's main claim to fame is to have raced many Derby and Oaks winners that became influential names in modern pedigrees. Nasrullah and Mahmoud are his greatest achievements, and Eclair is ancestress of Blushing Groom. Here is Nasrullah's pedigree:

				Polymelus
			Phalaris	**Bromus** ***by Sainfoin***
		Pharos		**Chaucer by** ***St. Simon-Canterbury Pilgrim***
			Scapa Flow	**Anchora**
	Nearco			**Rabelais** ***by St. Simon by Galopin***
			Havresac II	**Hors Concours (to** ***Galopin/St.Angela)***
		Nogara		**Spearmint**
			Catnip	**Sibola**
NASRULLAH				**Swynford** ***ex Canterbury Pilgrim***
			Blandford	**Blanche (carries** ***Galopin)***
		Blenheim		**Charles O'Malley** ***by son of St.Simon***
			Malva	**Wild Arum**
	Mumtaz Begum			**Roi Herode**
			The Tetrarch	**Vahren**
		Mumtaz Mahal		**Sundridge** ***ex Sierra (sister to Sainfoin)***
			Lady Josephine	**Americus Girl**

Leading sire in England and the USA, Nasrullah was English Champion 2YO colt. From 5 wins he won the Champion Stakes, Ascot Coventry Stakes, 2nd Middle Park Stakes, 3rd English Derby. He is linebred 5 x 5 to Sainfoin and his sister Sierra; 5 x 5 to Canterbury Pilgrim; and carries many sources of Galopin and his son St. Simon (including St. Simon's full sister via Havresac II).

WILLIAM HALL WALKER (Lord Wavertree) was a great breeder just before the year 1900 but I can't leave him out. He upgraded female lines that were acquired by other successful breeders. For example he bred Minoru (English Derby winner), Prince Palatine (11 wins including the English St. Leger, Doncaster Cup, Ascot Gold Cup) and developed the female lines responsible for Hyperion, Blandford and Princequillo. He sold Gondolette to Lord Derby, and developed the family of Black

Duchess via her daughter Black Cherry, dam of Cherry Lass, winner of the English Oaks and 1,000 Guineas. Cherry Lass's descendants include Night Hawk (English St. Leger), Blanche, Blandford, Sun Chariot, Santa Claus (English Derby), Carrozza (Oaks), Caro, Bebe Grande, Pieces of Eight, and Endear. Hall Walker also developed the female line of Cosquilla, dam of Princequillo. He in fact bred two ancestors which appear in Princequillo's pedigree, i.e. Minoru and Prince Palatine.

William Hall Walker owned the Tully Stud in Ireland (now the Irish National Stud) and was one of the most disciplined horse-breeders I've ever researched. He was especially critical of soft horses and only interested in breeding for soundness. His success mainly came from duplicating full brothers and sisters in the same pedigree of sound stallions. It was William Hall Walker who was directly responsible for encouraging the Aga Khan to venture into horses.

H.M. QUEEN ELIZABETH II & KING GEORGE VI - Queen Elizabeth II was leading breeder in England. The Queen, her father and the National Stud are responsible for Pall Mall, Aureole, Atlas, Height of Fashion, Highclere, Highlight, Angelola, Carrozza, Sun Chariot, Landau, Hypericum, Chamossaire, Big Game, Myrobella, Diligence, Agreement, Christchurch, Pindari, Rising Light, Milford, Dunfermline and Capo di Monte. Her Majesty's filly Height of Fashion became the dam of Nashwan, Unfuwain and Mukddaam. Landau was a leading sire in Australia.

The female line of Nashwan traces to Highlight, inbred 3 x 2 to Hyperion. She was the result of an inbreeding experiment created by King George VI, an admirer of Hyperion. Probably the intensely inbred Highlight inherited homozygous genes from Hyperion's influence.

LORD DEWAR - breeder of Tudor Minstrel, Fair Trial, Sansonnet, Cameronian, Commotion, Faux Tirage, Combat, Aristophanes, The Recorder, Abbot's Speed, Festoon, Medieval Knight, Ommeyad and Neolight. Lord Dewar was fortunate enough to acquire his foundation mare Lady Juror (Son-in-Law - Lady Josephine by Sundridge) which became the backbone of his breeding program. Tudor Minstrel was Champion 2YO colt and Champion Miler winning 8 races including the 2,000 Guineas, St. James's Palace Stakes, Knights Royal Stakes, Ascot Coventry Stakes and was regarded as one of the most brilliant athletes of the 20th century. He was given an extremely high rating by "Timeform". Fair Trial is of course the sire of Petition and Court Martial. Aristophanes sired Forli. Cameronian won the English Derby and 2,000 Guineas. Lady Juror has become one of the most important mares in pedigrees of stallions, and many Group winners receive impact from her son Fair Trial, and her daughter Sansonnet. Lady Juror was half-sister to Mumtaz Mahal.

MR. J. B. JOEL - breeder of Jest, Humorist, Pommern, Black Jester, Picture Play, Red Gauntlet, Pondoland, North Star III, Absurd, Polyphontes, Princess Dorrie, Major Portion. From a relatively small number of mares Mr. J. B. Joel established a successful branch of the Number 1 Family using Absurdity as the taproot. He developed this female line through Oaks winner Picture Play, ancestress of Chandelier, Crocket. Pommern and Humorist won the English Derby. Mr. Joel's main methods were to inbreed to superior sires (essentially Galopin and St. Simon) and to mix brother and sister, Sainfoin and Sierra, in the same pedigree.

THE EARL OF ROSEBERY - breeder of Blue Peter, Full Sail, Ribbon, Ocean Swell, Prue, Vaucluse, Hippius, Bongrace, and Afterthought. He practiced outbreeding rather than inbreeding and his major stakeswinners have not been influential in the long term, excepting for English Derby winner Blue Peter and his more influential brother Full Sail, leading sire in South America.

LORD ASTOR - breeder of Craig an Eran, St.Germans, Buchan, Rhodes Scholar, Sunny Jane, Short Story, Traffic Light, Saucy Sue, Book Law, Saltash, Trois Temps, Pennycomequick, Magpie, Pay Up, and Canon Law. The strength of Lord Astor's program was based on two foundation mares

Chelandry, from the Illuminata branch of the Number 1 Family, and Sceptre. Lord Astor bred an incredible number of stakes winning fillies descending from triple crown winner Sceptre. He was a great admirer of Derby winner Sainfoin and Derby winner Sunstar and considered them to be ideal sources for speed.

FRANCE:

MARCEL BOUSSAC - leading owner-breeder in France for many years. Breeder of Tourbillon, Corejada, Coaraze, Corrida, Djebel, Abjer, Djeddah, Pharis II, Galcador, Asmena, Acamas, Tourzima, Goya II, Coronation V, Appolonia, Adargatis, La Troienne, Ardan, Asmena, Crepellana, Abdos, Labus, Astronomie, Altana, Caracalla II, Arbar, Abdos, Auriban, Talma II, Coronis, Goyescas, Argur, Ambiorix, Emperor, Carissima, Only One, Furban, Pretty Lady, Scratch, La Moquese, Nirgal, Marsyas, Delsy, Cillas, Licata, Auriban, Thor, Elpenor, Akarad and Philius. One of the greatest breeders of classic winners of all time, Marcel Boussac experimented with close inbreeding, especially to his French Derby winner Tourbillon.

Prix de l'Arc de Triomphe winner Coronation V was a champion filly by a son of Tourbillon out of a daughter of Tourbillon. Marcel Boussac won more classics than any other European in thoroughbred history. For him, Pharis II became an influential son of Pharos. Tourbillon's son Djebel led the sires' list in France. While French Oaks winner Adargatis (by Teddy's son Asterus) was retained, her 3/4 sister La Troienne (by Teddy) was sold to Colonel E. R. Bradley's Idle Hour farm where she established a dynasty of stakes winners. La Troienne's descendants include Buckpasser, Sea Hero, Go For Gin, Caerleon, Striking, Straight Deal, Allez France, Admiring and Busher. An example of Boussac's inbreeding methods shows he was not afraid to experiment:

				Chouberski
			Bruleur	Basse Terre *by Omnium II to Pocahontas*
		Ksar		*Omnium II to Pocahontas*
			Kizil Kourgan	Kasbah *to Pocahontas*
	Tourbillon			*Rabelais*
			Durbar II	*Armenia*
		*DURBAN**		*Irish Lad*
			Banshee	*Frizette*
TOURZIMA				Ajax
			Teddy	Rondeau
		Asterus		Verdun
			Astrella	St. Astra
	Djezima			*Rabelais*
			Durbar II	*Armenia*
		*HELDIFANN**		*Irish Lad*
			Banshee	*Frizette*

Durban and Heldifann are full sisters. Tourzima is inbred 2 x 2 to these siblings and Ksar is inbred 3 x 2 to the stayer Omnium II. Tourzima produced Corejada (1st Irish Oaks, Cheveley Park Stakes, French 1,000 Guineas, Lowther Stakes, 2nd French Oaks, 3rd Yorkshire Oaks), herself dam of Apollonia (1st French Oaks, Grand Criterium, French 1,000 Guineas, Prix Morny) and Macip (1st Ascot Gold Cup, Prix Royal Oak). Tourzima also produced Gloriana (ancestress of Akarad, Akiyda, Licata) and Albanilla, fourth dam of Darshaan - 8 wins, Champion 3YO colt in France, 1st French Derby Gr-1, Prix Greffulhe Gr-2, Prix Hocquart Gr-2, Criterium de Saint-Cloud Gr-2.

Darshaan's best progeny to date are Kotishaan, Mark of Esteem, Truly a Dream, Darnay, Aliysa, Grand Plaisir, Sueborg, Game Plan and Hellenic. Thus, the genetic impact from the inbred Tourzima has lasted at least six generations!

FRANCOIS DUPRE - responsible for champion Tantieme, winner of the Prix de l'Arc de Triomphe Gr-1 twice, Rheffic, Sega Ville, Koblenza, Pola Bella, Reliance (French Derby), Relko (English Derby), Mirna, Texanita, Dushka (French Oaks), Tanerko, Polic, Bella Paola, Northern Light, Texana, Match III, and Fortino. Most of the Dupre horses were outbred (outcrossed individuals with no linebreedings). The male lines of Tantieme, Relko, Reliance and Rheffic are in serious trouble. Although these were high-class performers, they transmitted wide genetic variation and lacked prepotency. Thus, the Dupre bloodstock has too much variety in its make-up to provide dominance. However, two female lines have been developed from the Dupre program, that of Bella Paola and Texana.

ADARGATIS (Asterus-Helene de Troie) - Marcel Boussac's winner of the Prix de Diane Gr-1 (French Oaks) which is 3/4 sister to La Troienne. She produced stakes winners Pardal and Gabador.

ITALY:

FEDERICO TESIO - breeder of unbeaten Nearco (14 wins in France and Italy including the Grand Prix de Paris, Gran Premio di Milano, Premio Parioli, Premio Principe Filiberto, Premio Chiusura, Criterium Nazionale and the Grand Criterium), unbeaten Ribot (16 wins including the Prix de l'Arc de Triomphe twice, King George VI & Queen Elizabeth Stakes, Grand Criterium, Premio Emanuele Filiberto, Gran Premio del Jockey Club and Grand Premio di Milano), and Niccolo dell'Arca (Champion 2YO an 3YO in Italy, won Italian Derby). Other classic winners and Group performers include Antonio Canale, El Greco, Ruysdael, Boticelli, Brueghel, Tissot, Torbido, Astolfina, Cavaliere d'Arpino, Bellini, Theodorica, Tenerani, Delleana, Marguerite Vernaut, Tokamura, Toulouse Lautrec, Daumier, Duccio, Nera di Bicci, Nakamuro, Nervesa, Romanella and champion stayer Donatello II.

Owner of Dormello Stud situated in northern Italy, Federico Tesio is regarded as the most famous racehorse breeder of all time. He bred for the racetrack, trained all his great horses with the exception of Ribot, and created some of the best sire material the world has ever seen. Nearco became a legend. Ribot is still regarded by many as the best racehorse on turf. Donatello II sired Alycidon and Crepello. Tesio sought high-class female lines but didn't want to outlay large sums of money. He would buy fillies of moderate racetrack ability that were closely related to tough, sound stayers of classic-winning ability. He purchased Catnip because she was by English Derby winner Spearmint from English 1,000 Guineas winner Sibola. Catnip was a very plain filly and didn't fetch much at Tattersalls. Tesio overlooked her phenotype, relying on her "family" to make her a successful producer.

Tesio used all three methods of *inbreeding, linebreeding and outcrossing* to breed his classic winners. His best performers were linebred to *sons and daughters* of English Derby winners or leading sires. His home-bred pedigrees reflect a preference for St. Simon (Galopin-St. Angela by King Tom) and his sister Angelica; also classic winners Tracery (Rock Sand-Topiary by Orme) and Spearmint (Carbine-Maid of the Mint by Minting). He would linebreed to special mares to reinforce maternal strength. Where possible, he would take advantage of any proven "nick" that might give him a superior runner, and generally bred horses for the classic distances of a mile to a mile and a half. He was not interested in training sprinters. To his credit, Tesio wasn't afraid to use English Derby winners unpatronized by commercial breeders. In fact, he bred their best sons. However, like William Hall Walker, he would not tolerate unsound racehorses for his breeding program.

Nearco's male line continues via Nearctic, Dante, Nasrullah, Royal Charger - from which we get Northern Dancer, Bold Ruler, Seattle Slew, Halo, Turn-To, Storm Bird, Derring-Do, Never Bend, Roberto, Danzig, Lyphard, Sadler's Wells, Danehill and Sunday Silence. Ribot's line comes via Graustark, His Majesty, Tom Rolfe (sire of champion Hoist the Flag), Arts and Letters, Ribero, Con Brio etc. Probably the strongest male line of Ribot exists through his grandson Pleasant Colony, US Champion 3YO colt, Kentucky Derby winner and leading sire, although there are other highly bred sons of His Majesty young enough to promote the line next century. Graustark is the sire of Key to the Mint, a horse with a "triple" of Tracery.

NORTH AMERICA:

James Keene and Major Daingerfield were outstanding breeders of the American racehorse. They established domestic strains quite different to the fashionable imported English strains which came to Virginia, Kentucky, New York and Maryland. In the first half of the 20th century, the standouts among thoroughbred breeders responsible for upgrading the American product to international performance levels were:

AUGUST BELMONT II - breeder of Fair Play, Man O'War, Tracery, Trap Rock, My Play, Mad Hatter and a host of stakes winning fillies, sold his best yearling to Sam Riddle. Man O'War was champion at two and three years and for Sam Riddle became leading sire and leading broodmare sire. His champion son War Admiral figures as one of all time great broodmare sires in North America. August Belmont II discovered the "nick" of putting Rock Sand's daughters with Fair Play and numerous stakes winners resulted for him and his clients.

Tracery became leading sire in Argentina after siring a couple of crops in England. He is a vital element in the genotypes of modern Group 1 and Grade 1 winners and is found in the genetic make-up of four significant sires, viz. Nearctic, Alibhai, Ribot and Princequillo.

Tracery not only carries inbreeding to St. Simon and his sister Angelica, but also possesses the influence of Wellingtonia, the French stallion which gets impact from two daughters of *Double copy mare* Pocahontas. Could this be the main reason why Federico Tesio was intent on acquiring Tracery's strain for his breeding program? Barbara Burrini (by Tracery's son Papyrus) was bought at auction by Tesio and she became the dam of Italian champion 2YO filly Romanella which produced Ribot.

HARRY PAYNE WHITNEY - breeder of Whisk Broom II, Chicle, Equipoise, Dis Donc, Traverse (dam of Traffic), Pennant, Regret (first filly to win the Kentucky Derby), Mother Goose (ancestress of Northern Dancer and Halo), Whiskery, Whichone, Boojum, Nightstick, Burgomaster, Johren, Rosie O'Grady, Tryster, Bunting, Tanya, Artful, Fly Swatter and Top Flight.

He bred 192 stakes winners. The crossing of Broomstick's strain with that of Peter Pan was a mix which upgraded speed and Harry Whitney made full use of this "nick". Dis Donc, an unbeaten two year old, was bred this way. The fast filly Fly Swatter was linebred to famous mare Mannie Grey three times and produced 1946 champion 2YO filly First Flight. Frilette (Broomstick-Petticoat) was a valuable foundation mare for Harry Whitney and is ancestress of Silver Fog and Silver Spoon.

Harry's son Cornelius Vanderbilt Whitney developed a stakes winning line from Mother Goose, the grand-dam of the elite mare Almahmoud. Harry's eldest brother Payne established Greentree Farm and bred Goose Egg, dam of Kentucky Derby winner Shut Out.

JOHN HAY WHITNEY - with his sister, established Greentree Farm into one of the most successful nurseries in Kentucky and purchased Tom Fool - US Horse of the Year for stallion duty. John Hay Whitney bred Stage Door Johnny, The Axe II, Capot, Hall of Fame, Stop the Music, Bowl Game, Late Bloomer, Bebopper, etc.

COLONEL E. R. BRADLEY - owner of Idle Hour Farm, stood Blue Larkspur and imported La Troienne - Marcel Boussac's four year old mare that was culled and sold at Newmarket. Colonel Bradley bred Bimelech, Black Servant, Bubbling Over, Balladier, Behave Yourself, Businesslike, Baby League, Black Helen, Broker's Tip, Boot to Boot, Black Gold, Black Maria, Big Hurry, Burgoo King, Bazaar, Belle of Troy, Busher, Be Like Mom, But Why Not, Bridal Flower, etc.

La Troienne is one of the most influential mares in the American stud book. She was by Teddy out of Helene de Troie by Helicon from Lady of Pedigree by St. Denis and came from a branch of the Number 1 Family. In phenotype, La Troienne was a small, plain mare, quite unlike her three-parts sister Adargatis which won the Prix de Diane (French Oaks) but she would transmit tremendous speed to her progeny and to their descendants.

Her sire Teddy was inbred to Ormonde and Doremi, genetic 3/4 brother and sister, both being by

Bend Or from daughters of Macaroni. La Troienne's dam Helene de Troie was by Helicon, a grandson of Bona Vista which was also bred on the Bend Or/Macaroni "nick". Teddy's sire was unbeaten French classic winner Ajax, whose dam was linebred to the *Double Copy mare* Pocahontas. Ajax was out of Amie by Clamart (traces to Pocahontas) from Alice by Wellingtonia (carries two daughters of Pocahontas). La Troienne's grand-dam carried St. Simon, whose broodmare sire King Tom was a "son" of Pocahontas.

Blue Larkspur's male line is almost non-existent. He was an exceptionally dominant sire of fillies, stamping his phenotype on this sex, but not so on his colts. Blue Swords was his best son at stud and he is more known as a broodmare sire. Better Self, from the same line, proved a great sire of fillies.

EDWARD P. TAYLOR - Canada's premier owner -breeder. Eddy Taylor was a dynamic business-man who established Windfields Farm in Ontario. His forte was to buy well-performed fillies off the racetrack as well as mares from strong English female lines. He stood many successful sires at his farm including Chop Chop, Windfields, Bunty Lawless, Bull Page. He purchased Lady Angela (Hyperion-Sister Sarah) from Newmarket's Tattersalls sale because he felt she might make a foundation mare. She was tested in foal to Nearco, and he arranged for the mare to revisit the same stallion before shipping her to Canada. There she produced Nearctic, winner of 21 races $152,354 and voted Horse of the Year. He won the Saratoga Special, Victoria Stakes, Clarendon Stakes, Carleton Stakes, Swynford Stakes, Jacques Cartier Stakes, International Handicap, Bold Venture Handicap and was leading sire in Canada seven times.

E. P. Taylor bred Nearctic's best son Northern Dancer, 14 wins from 18 starts $580,647 including the Kentucky Derby, Preakness Stakes, Flamingo Stakes, Remsen Stakes and Queen's Plate. Northern Dancer was shipped to Windfields Farm in Maryland when he shot to the top of the sires' list. E. P. Taylor holds the record for breeding more stakes winners than anyone else during the 20th century. Taylor's main theme was to buy into strong female lines and match them with the very best stallions money could buy.

WARREN WRIGHT - owner of Calumet Farm. One of the leading breeders for two decades, Warren Wright bred many stakes winners including Whirlaway, Citation, Twilight Tear, Columbiana, Sun Again, Bull Lea, Coaltown, and a host of good fillies.

ALFRED G. VANDERBILT - breeder of Champion Native Dancer, only once defeated in a brilliant racing career. He purchased Discovery in training and made him into a champion sire. Vanderbilt bred Miss Disco, dam of America's leading sire Bold Ruler.

Other prominent breeders were Joe E. Widener, Arthur B. Hancock, William Woodward, C. T. Chenery, Hal Price Headley and Louis B. Mayer. Elmendorf Farm, the oldest thoroughbred nursery in Lexington, was also a major contributor and bred Preakness Stakes winner Polynesian, sire of Native Dancer.

We now come to the second half of the 20th century.

ENGLAND & IRELAND

If one studies the breeders of classic winners between 1951 to 1996, no one breeder had a stranglehold on the best breeding stock in the British Isles. Many outstanding mares were sold by British and Irish breeders to American buyers after 1945. The exodus of classic-winning colts to the USA had its effect on the improvement of the English thoroughbred.

However, among the most successful breeders of stakes winners were Lord Howard de Walden, Robert Sangster, His Highness Karim Aga Khan, H. J. Joel, Major L. B. Holliday, Miss Dorothy Paget, Lord Rosebery, His Highness Sheikh Maktoum bin Rashid al-Maktoum and his brothers Sheikh Hamdan and Sheikh Mohammed, Prince Khalid Abdullah, Captain D. Rogers, Paul Mellon, W. and R. Barnett, G. A. Oldham, Sir John Astor, R. A. Budgett, Lord Halifax, Lord Irwin, Major J. H. de Burgh, George Strawbridge Jr., Eve Stud Limited, Banstead Manor Stud, Joseph McGrath, Overbury Stud, Ballymacoll Stud, Coolmore Stud, Moyglare Stud, Cheveley Park Stud, Snailwell Stud, Airlrie Stud and White Lodge Stud.

FRANCE:

French breeding has been stifled by its Government which does not offer enough incentives to compete favorably with England or Ireland. Leading breeders in France include Stavros Niarchos, Baron Guy de Rothschild, Comte Roland de Chambure, Alec Head, Daniel and Alec Wildenstein, Countess Margit Batthyany, Jacques Wertheimer, Francois Dupre, Madame L. Volterra, Madame Arpad Plesch, Madame J. Couturie, Jean Ternynck, Sir Philip Oppenheimer and Paul de Moussac.

NORTH AMERICA:

There are many successful breeders in the USA and Canada, but the short-list comprises Arthur "Bull" Hancock, Warner L. Jones, Jr., Will S. Farish III, W. S. Kilroy, Josephine Abercrombie, Alice Chandler, Allen Paulson, Leslie Coombs, Ogden Phipps, F. W. Hooper, Captain Harry F. Guggenheim, Nelson Bunker Hunt, Arthur B. Hancock III, Seth Hancock, Darby Dan Farm, Beiber-Jacobs Stable, John Gaines, Ethel D. Jacobs and Harbor View Farm, Ogden Mills Phipps, William T. Young, Robert Clay, Robert E. Meyerhoff, John and Betty Mabee, Jack Kent Cooke, Live Oak Stud, Derry Meeting Farm, Tom Gentry, Tartan Farms, Runnymede Farm, Leone J. Peters, Mr. and Mrs. Paul Hexter, Helen C. Alexander, Graham Beck, Edward P. Evans, North Ridge Farm, Brereton C. Jones, Charles Taylor, Cardiff Farm, Peter M. Brant, Charles and Alfred Nuckolls, Fares Farm, Prestonwood Farm, Calumet Farm and Pope McLean's Crestwood Farm.

Warner L. Jones, Jr. was the smartest commercial breeder I ever met. "Bull" Hancock was responsible for introducing syndication of stallions to North America. Ogden Phipps and his son Ogden Mills Phipps possessed the strongest broodmare band in the 1980's and bred many successful stallions. Ogden Mills "Dinny" Phipps, an investment banker, was Chairman of the Jockey Club and won the 1978 Eclipse Award of Merit. Unfortunately, American breeders sold too many quality mares and yearling fillies to Japanese and European buyers and the pendulum is swinging in different directions. The Japan Racing Association (JRA) restricts foreign competition. This ruling has met with worldwide criticism and one hopes the ban will be lifted.

JAPAN:

Zenya Yoshida established Shadai Farm in the north island of Hokkaido and for twenty years it has been a production house for Group 1 winners. Japan's leading sires Northern Taste and Real Shadai are high-class stallions and Mr. Yoshida's sons continue to buy some of America's best fillies and mares to strengthen their hold as premier breeding establishment in Japan. They stand the world's leading sire of 1995, Sunday Silence, an almost black son of Halo.

AUSTRALIA & NEW ZEALAND:

Among the most professional farms in Australia are Arrowfield Stud, Widden Stud, Woodlands Stud, Segenhoe Stud, Newhaven Park Stud, Muranna Stud, Collingrove Stud, Lindsay Park Stud,

Baramul Stud, Lyndhurst Stud and Emirates Park Stud. The brothers Jack and Bob Ingham have developed Woodlands Stud into a magnificent nursery and stand Australia's champion 2Y0 colt Canny Lad (by Bletchingly) plus shuttle stallions Brief Truce, Quest For Fame and Grand Lodge.

Arrowfield Stud is one of the largest consignors of yearlings and one of the most successful breeders of Group 1 winners. Up until 1995, Danehill stood at Arrowfield. He is now leading sire in the country. Danehill's son Danzero, winner of the $2 million STC Golden Slipper Stakes Gr-1 is now at stud and stands alongside shuttle stallions Catrail, Brocco and Chief's Crown. A public company, Arrowfield Stud is based at Scone in the rich Hunter Valley and its chairman is John Messara. Many imported mares grace the paddocks at Arrowfield.

Segenhoe Stud is also a public company and a leading consignor of yearlings, It is under the direction of Michael Sissian and Tony Bott. Based at Scone, Segenhoe Stud stands Palace Music (sire of Cigar) and his son Naturalism, Australia's Horse of the Year and Champion 3YO colt. Widden Stud is home of Marscay, an influential sire which descends in male line from Star Kingdom. Champions have been bred at this farm which stands the shuttle stallions Barathea, Mukaddamah and Johann Quatz as well as Breeders' Cup Juvenile Stakes Gr-1 winner Success Express (by Hold Your Peace) and Graded stakes winner Don't Say Halo (by Halo). Robert Sangster in partnership with Colin Hayes own Collingrove Studs, based in two states. They stand Bluebird, Turtle Island, Scenic and Golden Slipper Stakes Gr-1 winner Rory's Jester.

In New Zealand, Cambridge Stud is home of premier sire Sir Tristram and of his Group winning son Zabeel (dam by Nureyev). This commercial operation is run by its owner Patrick Hogan and is responsible for breeding many champions including the current sensation Octagonal.

Allen Paulson, a leading owner-breeder in USA who races Cigar. the world's record stakes earner. Paulson owns Brookside Farm in Kentucky.

Edward P. Taylor, breeder of Kentucky Derby winner Northern Dancer.

ASIAN MARKETS AND THEIR IMPACT IN THE 21st CENTURY

"To be or not to be, that is the question?"
- William Shakespeare

The economic might of Asia is already being felt in countries like Australia and New Zealand where overseas buyers take more than 40% share of total spending for yearlings at the Sydney and Melbourne sales. The Korean Jockey Club already encourages the importation of breeding stock and Korean investors should be major players at Tattersalls and Keeneland in the next decade.

The People's Republic of China plan to establish seven new racetracks on mainland China, and of course, Britain's lease of Hong Kong expires in 1997. Already there have been overtures made by Chinese Government officials about the possibility of developing thoroughbred farms. Until now, racing in China has been restricted to competition amongst Mongolian ponies. The amount of gambling on these ponies is mind-boggling enough for the Government to spot another source of revenue. Hong Kong's daily turnover is staggering!

Australia and New Zealand sit on the doorstep of Asia. Therefore it is justifiable for Australasian thoroughbred breeders to invest more heavily in quality stallions and mares so as to compete with England, Ireland, and North America and gain the biggest slice of the Asian market. An inaugural yearling sale in Hong Kong was very successful. It is to be repeated. The history of the Australian thoroughbred commences shortly after 1788 when an English captain became Governor of the new colony at Sydney Cove. Governor Phillip encouraged the importation of thoroughbreds, sheep and cattle and by 1800 amateur races were held. No nation loves horse-racing more than Australians. Simulcasting of racing in five states increases betting turnover. All racing is conducted on turf courses and because of Australia's warm climate in summer and autumn, fast tracks test the soundness of thoroughbreds. The six most important Group 1 races are:

1. **The $2 million (plus) W. S. Cox Plate Gr-1 (standard weight-for-age) over 2,000 metres (1 1/4m) at Moonee Valley racecourse is the annual event which can determine Australia's Horse-of-the-Year title.**

2. **The $2 million Caulfield Cup Gr-1 over 2400m, is the nation's most prestigious handicap over a mile and a half.**

3. **The $2.5 million Melbourne Cup Gr-1 over 3,200m is the nation's two miles handicap, and has formed part of Australia's culture. It attracts international stayers.**

4. **The $2 million (plus) Sydney Turf Club's Golden Slipper Stakes Gr-1 is the richest race for two year old colts and fillies going 1200m at weight-for-age.**

5. **The Australian Jockey Club Derby Gr-1 over 2,400m at Randwick racecourse is the nation's most important classic for three year olds.**

6. **The Victoria Derby Gr-1 over 2,400m at Flemington racecourse for three year olds.**

The winner of any of these Group races has to be of international standard and it is surprising that the winners of the Caulfield Cup, AJC Derby and W.S.Cox Plate are not sent overseas to compete in the Prix de l'Arc de Triomphe Gr-1 or Breeders' Cup Classic Gr-1. Having lived in the United States and Australia, experienced racing there and in other countries, I believe I can impartially judge whether Australian stakes winners would be competitive elsewhere. I have no hesitation in stating that Australia's best sprinter-milers are world class.

New Zealand-bred Balmerino was runner-up in the Prix de l'Arc de Triomphe Gr-1 at Longchamp; Horlicks won the Japan Cup Gr-1; and Australia's Better Loosen Up won the Japan Cup Gr-1. Those of you old enough might remember when Phar Lap scored an amazing win in the Agua Caliente Handicap defeating some of the best horses in training in California.

Australians do not tolerate "bleeders" which are banned for three months. If a horse bleeds again, it is banned for life. There is an increasing number of bleeders being produced in the northern hemisphere and because Australia does not have the problem in huge numbers, many of its Group winners should be imported to England and the United States of America.

In my opinion, there is potential for northern hemisphere breeders to buy Australian mares and fillies which posssess different genetic strains to fashionable strains of Europe and North America. The quality of bloodlines in Australia has never been better. South African buyers come to the major yearling sales held in Sydney and Melbourne and compete for the best colts and fillies against Asian buyers.

In the 21st century I can see more Australian horses performing around the world. Nor would it surprise me to learn that Japanese breeders will purchase horse farms in Australia once the JRA ban has been lifted. They are quite familiar with Australian pedigrees and regularly attend our sales. After all, it is a land of sunshine, with beautiful green golf courses, the best surfing beaches in the world, a natural wonder called the Great Barrier Reef and unrivalled scenery that varies from the tropics in Queensland to the cooler climate of Tasmania. The shuttling of Group 1 stallions to Australia and New Zealand have brought the standard of racing to its highest level. Next century the Australian-bred and New Zealand-bred stallion should find its way to Newmarket and Kentucky.

SOUTH AFRICA

In a discussion I had recently with Gainesway Farm's managing director Antony Beck about how South African racing is progressing he very kindly gave me a quick overview. He advised the yearling market is becoming stronger and that by the turn of the century it should be more than solid. Young stallions in South Africa which are making an impact as sires of Group winners include National Assembly (by Danzig), Al Mufti (by Roberto), Model Man (by Elliodor, son of Lyphard from the family of Busted), and American stakes winner Badger Land. Among influential sires on the sires' list, mention should be made of the dominant Foveros (by Averof), as well as popular commercial stallions such as Harry Hotspur (by Mexico II), Thatching's full brother Golden Thatch (by Thatch), and Northern Guest, the full brother to classic winner El Gran Senor.

Antony's father Graham J. Beck (son of a stockbroker) was educated at Kingswood College in Grahamstown and the University of Cape Town, South Africa and owns Maine Chance Stud. Part owner of Timber Country - U.S. Champion 2YO colt of 1994 - he purchased Gainesway Farm in Kentucky in 1989. In South Africa he has headed the leading owner and breeder's list many times. Graham Beck purchased champion 3YO filly Winning Colors (Kentucky Derby Gr-1 winner) to add to his quality broodmare band.

CHAMPIONS OF THE CENTURY

"Courage overcomes fear."
- Alexander the Great

In concluding "Genetic Heritage" which I trust is a helpful overview of the development of the modern racehorse, I would like "Champions" of the 20th Century to be recalled. We have seen many great horsemen in the saddle at Royal Ascot, Epsom, Newmarket, The Curragh, Longchamp, Churchill Downs and Randwick. My "Champions" to make the honor roll were selected on merit. No doubt the reader will share differences of opinion. The best European jockey of the century is Lester Piggott. The best North American jockey is Bill Shoemaker. The most talented jockey seen in Australia is the incomparable George Moore. These three riders judged pace, used skillful tactics and rode with consistency: each had a gentle pair of hands and master-minded victory. The Heritage Awards for the Champions of the 20th Century are:

The most influential sire:
NEARCO (Pharos-Nogara by Havresac II) unbeaten champion racehorse bred, owned and trained by Federico Tesio. Nearco established the strongest male line of the century and transmitted clean air-flow. He possessed no genetic weaknesses and upgraded for soundness and speed.

The most influential mare:
MUMTAZ MAHAL (The Tetrarch-Lady Josephine by Sundridge) Champion English 2YO filly bred by Sledmere Stud and purchased by His Highness The Aga Khan. She established a dynasty and her male descendants became influential sires thereby gaining her my vote over other mares such as Selene, Plucky Liege and La Troienne.

The most brilliant sprinter-miler on Turf:
Best Male - TUDOR MINSTREL (Owen Tudor-Sansonnet by Sansovino) Champion English 2YO colt, Champion Miler, winner of the English 2,000 Guineas, bred by Mr. J. A. Dewar, wins the title for best male. Best Female - MIESQUE (Nureyev-Pasadoble by Prove Out) multiple champion on grass, champion European filly at 2 and 3 years and twice winner of the Breeders' Cup Mile Gr-1. She was bred by Stavros Niarchos.

The most brilliant sprinter-miler on Dirt:
Best Male - HOLY BULL (Great Above-Sharon Brown by Al Hattab) bred by Pelican Stable in Florida. He gets my vote over Dr. Fager. Best Female - RUFFIAN (Reviewer-Shenanigans by Native Dancer) a true champion who should be remembered.

Champion Turf Horse:
Best Male - RIBOT (Tenerani-Romanella by El Greco) undefeated, twice winner of the Prix de l'Arc de Triomphe, and of the King George VI and Queen Elizabeth Stakes. He was bred by Federico Tesio and the Marchisa della Rochetta. Best Females - DAHLIA and ALLEZ FRANCE.

Champion Dirt Horse:
Best Male - SWAPS (Khaled-Iron Reward by Beau Pere) who set many world records and was bred by Rex Ellsworth. I rate him superior to Buckpasser, Man O'War, Secretariat and Cigar. Best Female - LADY'S SECRET champion daughter of Secretariat, and toughest mare seen in the USA.

Champion 2YO turf::
Best Male - ARAZI (Blushing Groom-Danseur Fabuleux by Northern Dancer)
Best Female - MUMTAZ MAHAL (The Tetrarch-Lady Josephine by Sundridge)

Champion 2YO dirt:
Best Male - SECRETARIAT (Bold Ruler-Somethingroyal by Princequillo)
Best Female - FLANDERS (Seeking the Gold- Starlet Storm by Storm Bird)

Of all the horses to have raced in the southern hemisphere, one stands out above all others and his name is TULLOCH, son of the Aga Khan-bred stallion Khorassan. Tulloch was foaled and raised in New Zealand, trained by T. J. Smith and set track records like no other horse before or since. He was voted Horse of the World for his generation.

Champion Stayer:
PHAR LAP (Night Raid-Entreaty), foaled in New Zealand and a horse with a mighty big heart.

The most influential broodmare sire:
CHAUCER (St. Simon-Canterbury Pilgrim by Tristan). He was bred by Lord Derby in England.

Trainer's Award:
Europe: Irish trainer Vincent O'Brien, a remarkable horseman.
North America: Ben Jones and D. Wayne Lukas (equal). Ben Jones was capable of getting into the mind of a horse and trained for Calumet Farm. Wayne Lukas has the record on the board.

Breeder's Award:
Federico Tesio, a genius with the racehorse. He was called the "wizard" of Dormello.

SECRETARIAT wins the Kentucky Derby Gr-1 by two and a half lengths from Sham in race record time of 1-59 2/5 for 1 1/4 miles.

CONCLUSION

The thoroughbred is an athletic animal specifically bred for the racetrack. The selection process has been geared to increase speed over a distance. Colts and fillies born at the end of the 20th century are in fact superior to those born at the beginning of the century. Whether they are more physically sound is questionable, and herein lies my greatest concern. If we are unable to improve our selection skills and breed sound racehorses then trainers are entitled to grumble. One of the questions which puzzles me most is the dominance of Northern Dancer in the 1980's and 1990's. This stallion was not considered by North American breeders to be a sire of dirt runners, despite the fact his sons and grandsons are getting Graded stakes winners on dirt. In 1984 there were twelve Northern Dancer yearlings at Keeneland's select sales and they averaged $3,446,667 each - $41 million out of the entire $175 million! Windfields Farm took home $17.7 million for their consignment of fourteen yearlings. In a recent interview, successful trainer John Nerud criticized Northern Dancer for transmitting an element of unsoundness. I respect John Nerud (he managed Tartan Farms in Florida) because he is an experienced horseman. We should heed his opinion. He knows the stallion's stud career record and his success rate. I happen to share similar doubts about Northern Dancer with John Nerud. I've observed at first-hand, weaknesses prevalent in his progeny and certain things worry me about the long-term affects of duplicating Northern Dancer too close. Please consider these points:

(1) First, Northern Dancer hasn't yet made a huge impact as a broodmare sire. Very few of his daughters have produced Group 1 winning colts that became successful commercial sires. It will be interesting to watch Arazi's stud career because he is from a daughter of Northern Dancer; but unlike others out of Northern Dancer mares, Arazi is an inbred horse - inbred to the very influential French stallion Wild Risk - a tough horse.

(2) Second, Edward P. Taylor and his son Charles were aware that Northern Dancer was not above average in terms of his fertility which in normal circumstances might have been a commercial drawback. However, Eddy Taylor and his son Charles wisely restricted the number of mares Northern Dancer was allowed to cover in any year, using a stallion service agreement that limited bookings. Subsequently, we discovered a small percentage of Northern Dancer's sons would prove to be infertile. About 20% of Northern Dancer's sons had *below* average fertility. Therefore, if we are wanting to improve fertility, in my opinion, he is not a horse one would wish to duplicate in pedigrees within say, three generations of a foal's pedigree.

(3) Northern Dancer sired a percentage of "bleeders", particularly fillies - not on a grand scale, mind you, but enough to alert breeders that a recessive gene is present in his make-up. Duplicating Northern Dancer in pedigrees certainly will upgrade fast twitch muscle fiber - one of the stallion's best assets.

(4) Northern Dancer's strain is more successful via his sons than his daughters. Why?

I can understand John Nerud's reservations. However, Northern Dancer transmitted tremendous speed and this is why he became a leading sire. If one wants to duplicate Northern Dancer in a foal's make-up it should be accomplished via his soundest descendants. For example, I like the mix of Nijinsky II (not exactly a true Northern Dancer type) with Storm Bird. I believe it can work to produce a stakes runner if the duplication is not too close in a pedigree, preferring it to be no closer than the fourth generation, or further back. Storm Bird and Nijinsky both have dams with Bull Page's influence. They are almost genetic cousins.

Nureyev is a Northern Dancer horse with an unusual pedigree pattern and I can see his descendants

being successful when combined with certain other sound Northern Dancer lines with good respiratory systems. Nureyev happens to have genetic influence from the English sprinter Gold Bridge (carrying Orme's strain) and Rough Shod's extremely well-bred dam Dalmary, a mare which is closely related to Blenheim. Thus, when Nureyev's strain is matched with say a descendant of Royal Charger or Nasrullah which both carry Blenheim's input from Mumtaz Begum, the Blenheim link *with* Dalmary should be effective for speed. It should produce above average runners, maybe even stakes winners.

I can imagine success coming from a doubling of Northern Dancer if the mare Almahmoud is duplicated three times in a foal's pedigree. Almahmoud (Mahmoud-Arbitrator) was a very sound race mare even if she was hot-tempered. Bering, winner of the French Derby and runner-up in the Prix de l'Arc de Triomphe Gr-1 is a stallion possessing linebreeding to Almahmoud. He is by Sea Bird II's son Arctic Tern. If daughters of Nureyev are matched with him it would certainly double up on Northern Dancer fairly close, but more importantly, it would give a triple source of Almahmoud, she being by the very sound Mahmoud. I could live with this type of mating because Bering was a very sound racehorse.

What concerns me most is the saturation of Northern Dancer's strain in Europe. Almost 70% of commercial stallions in England and Ireland carry Northern Dancer's strain. If most horse-breeders have a high percentage of Northern Dancer appearing in their fillies and mares, it is inevitable that duplication of the stallion will eventuate. This means European breeders will find out soon enough about Northern Dancer's strengths and weaknesses. Northern Dancer's "recessive" genes will have an ideal chance to be expressed.

By using the soundest descendants of a great sire, one is being more *selective.* Any stallion or mare which is to be part of a plan to help "fix" desirable traits must be of the very highest quality and be really sound.

My next puzzle is how to design more competitive matings. If hundreds of foals with inbreeding to Northern Dancer are entered for races, how does one breed something faster? Logic tells us that if many different owners are racing stock with similar genes, there will be an increase in the number of photo finishes! An owner must therefore look either to use the absolute *best* descendants of Northern Dancer, or else choose to use *different* superior genetic material.

Where does the answer lie? Well, in the next decade, I'll observe the success rate of Turn-To's sire line for he has grandsons which are superior on the racetrack. I will study all top-class runners from the Damascus line, especially waiting for the first crops sired by Gilded Time and Technology. I will study all the best grandsons of Mr. Prospector, especially Fappiano's sons. Darshaan is another to observe. I will focus on Ribot's best sire son His Majesty, for here lies a very sound strain. Any one of these strains might provide the next great sire of Europe or North America.

History sure has a habit of repeating itself. In England at the beginning of the century, St. Simon had a multitude of sons and grandsons at stud and was just as popular as Northern Dancer is today. After two generations of dominance his male line went into low gear in England, yet survived in France where St. Simon's lesser son Rabelais was matched with *different strains.* My question is this: Had saturation point caused the decline of St. Simon's male line in England?

The rise and fall of sire lines is a mystery. Bold Ruler was the hottest sire in the world, yet how powerful is his male line today? Could anyone have guessed he would be so lightly represented by 1996? Sir Gallahad III and Bull Dog shared the spotlight in the 1940's and 1950's with many sons and grandsons given every opportunity to succeed as sires in Kentucky. Where is Sir Gallahad's

male line today? Where is it hiding? He had quality sons like Gallant Fox, Fighting Fox, Black Devil, Bold Irishman, Roman, Gallahadion and Hadagal to represent him, but their lines have faded. Why? Bull Dog was represented by sons like Bull Lea, Eternal Bull, Occupation, Occupy, Coldstream, Our Boots, The Doge and Tiger. Why did his male line decline?

Next we must ask ourselves how important is maternal strength. Great sire lines require maternal strength to survive, so is there some biochemical reason? Why do I advise people to research a stallion's dam? Because the dam's influence is critical. She makes the difference as to whether her son will make a successful sire. If this wasn't true, *every* well-performed son of a successful sire would make the General Sires' List! The bottom line is that only a rare few sons make commercial sires.

Northern Dancer's grandson Danehill (Danzig-Razyana by His Majesty) seems to be genetically "clean" and is siring many Group winners in Australia and Europe. He is 3 x 3 to Natalma which makes a difference and probably supplies prepotency. However, it is Danehill's broodmare sire His Majesty (linebred to Tracery) which seems to be the *main influencing factor for soundness and classic speed.* Danehill has several sons at stud, and two of the best are champion miler Flying Spur (winner of the STC Golden Slipper Stakes Gr-1, AJC All-Aged Stakes Gr-1, VATC Australian Guineas Gr-1) and Australia's millionaire champion 3YO colt Danewin which is out of a half sister to the sensational champion Octagonal.

Danewin (Danehill-Cothele House by Petition's son My Swanee) set a race record when he went a sub-two minute time to win the STC Rosehill Guineas Gr-1 over 1 1/4 miles and stands at H.E. Nasser Abdullah Hussain Lootah's Emirates Park Stud in New South Wales in 1996 at a fee of $20,000. Because of his versatility, speed and soundness, Danewin should make a first-class sire. He traces to Tessa Gillian, the sister of Royal Charger, and 3/4 sister to Nasrullah. It is the Number 9 Family of Mumtaz Mahal.

THE AIR-FLOW FACTORS

In 1993 a book titled "Specification for Speed in the Racehorse" written by Robert W. Cook, FR CVS PhD sub-titled "The Airflow Factors" appeared in bookshops. I enjoyed reading it. A few quotes are given below to tempt the reader to buy this interesting book:

"The evidence indicates that performance in all competitive horses is afforded by airflow. Racehorses with poor airflow are, in general, unable to run at top speed for long distances. Many breakdowns on the racetrack may trace their cause to fatigue brought on by difficulties in breathing. Racehorses are expected to work constantly at the peak of their potential and, because of this, small variations in airflow are reflected in small, but nevertheless, significant differences in performance as measured in fifths of a second."

"Since we have been unable to spot the potential for speed by viewing the horses from the outside, we have looked elsewhere. Speed and stamina are limited, I believe, more by the conformation of the horse's airway than by the length or straightness of its legs, the angle of its pasterns, or the slope of its shoulders. I find that only 5% of Thoroughbreds have wide and healthy airways. It is horses in this group that represent the best prospects of becoming stakes winners. Airflow factors can be measured in horses of one year old or older and all but one factor (height) can be measured in foals. Breeding for a wide and healthy airway may be the most effective way to improve performance. The airflow factors relate to fuel supply for the racehorse."

Robert Cook goes on to explain about the supply and demand for oxygen, and discusses recurrent

laryngeal neuropathy (RLN), a name he uses for a common disease of the nervous system. He states: "RLN may be inherited or acquired but 99% of cases are inherited. This disease may occur on the left side of the voice box, on the right side, or on both sides simultaneously. *Left-sided disease is by far the most common.*"

The recurrent laryngeal nerve is the most important nerve to the voice box and stimulates all the muscles of the voice box (except one), so any tampering of the nerve, or destruction of the nerve is of a serious nature. Many stallions and mares carry the "recessive gene" responsible for recurrent laryngeal neuropathy (RLN), and as a consequence, this disease spread during the 20th century. Thoroughbred breeders are to blame for the increase in the number of horses affected by this disease. Mares which displayed RLN should never have been used for breeding purposes. Unfortunately there is not enough discrimination in the mare population. A stallion which possesses a wide throat-latch might be able to pass this valuable trait on to its progeny. Selection for wide jaws makes a lot of sense. This feature helps improve airflow.

THE M AND X FACTORS

Now that we are learning more about the mitochondria of cells (the supplier of energy for living cells) and recognizing its unique *mitochondrial DNA properties,* should we respect strong female lines much more than we have in the past? Furthermore, if the X-factor for large heart size is sex-linked as we believe it is, shouldn't we first test young mares before culling them solely on the basis of performance? Might some of them inherit the XH chromosome as a recessive factor and therefore have a 50% chance of transmitting it to their foals?

Culling undesirable mares is extremely important. Those which transmit structural faults must go. They can't help produce you an athlete. Mares which produce more than one "wobbler" may inherit the weakness via recessive genes, indicating a congenital problem. I have a suspicion that a champion son of Native Dancer might be a major a source of this seemingly congenital malfunction. Sub-fertile mares shouldn't be tolerated. Mares with air-flow problems should never be allowed to produce a foal. In summary, breeders must learn to be more discriminate in the selection of females for production. If this philosophy were practiced, we'd rid the gene pool of inferior stock.

Fillies and mares with drug-free training careers usually make the best producers. "Chemically-aided" females are poor risks to make producers when they are retired from the racetrack. Drugs upset their metabolism. Isn't it odd that a lot of unraced mares make stakes producers?

An environment must be user-friendly since it affects the genotype. Young racehorses should be out and about, running with their pals for as long as possible to exercise their muscles. Physically fit yearlings are a pleasure to observe in large, open fields because this is the way God meant them to be raised. Yearlings must be allowed to run free because essentially they are a herd animal. By learning to run with each other they develop the spirit of competition. Why rob them of this right? (In other words, athletes should never be "hot-housed" or raised in small paddocks).

We should marvel at the thoroughbred. It is a remarkable piece of engineering and its genetic heritage traces from the beginning of time. The methods employed in the production of young racehorses is becoming more scientific with the aid of computerization and a knowledge of genetics. Next century we may see quantified scientific research. Whatever guided breeders in the past might not apply in the 21st century. DNA information and gene map data might help. Upgrading for speed and stamina requires some knowledge of family traits, otherwise how does one know how to upgrade? How does one select for improvement? Matings should be designed using information about conformation, fertility, temperament and airflow as well as performance.

The incidence of great mares in the pedigrees of leading sires must not go unnoticed. It is they which hold the key as to whether a male line might be taken one generation further. Inheritance comes from an equal distribution of both parent's genetic codes - transmitted by stallion and mare - yet because genes are either dominant, recessive, or sometimes co-dominant, a foal will inherit its genetic code in varying degrees according to the strongest influences - *the expression of dominant genes over recessive genes.* Nature will mask certain characteristics and the magic and mystery of thoroughbred breeding can continue to deliver unexpected results. Testing a special mating once may not supply the best foal. Therefore, a potentially good mating should be tested at least twice. There are many instances whereby the second foal of a repeat mating supplied a Group or Graded stakes winner.

PEAKS AND VALLEYS, Canada's Horse of the Year and Champion 3YO Colt being led back to the winner's circle by owner-breeder Josephine Abercrombie (Pin Oak Farm) after winning the Molson Million Gr-1 at Woodbine - Photo courtesy of Michael Burns, Toronto. This is a champion son of Mt. Livermore.

In 1996, Kentucky veterinarian Dr. Richard D. Holder correctly predicted the sex of 180 foals when they were only 60 to 70 days old in the utero of their dams. He has pioneered *fetal sexing.* I remember back in 1988 when Richard first began to test mares for fetal sexing at Overbrook Farm. Jim Cannon (Overbrook's farm manager) asked me to observe. In 1994, 70 out of 70 foals were sexed correctly by Dr. Holder. Now, breeders can discover the sex of a foal before it is born. This information is helpful when a breeder contemplates selling a mare at auction. Sometimes the sale of the mare can be postponed if one could be sure she carries a filly foal by a sire with *sex bias,* i.e. a sire of very good fillies.

Dr. Holder determines fetal sexing by locating the genital tubercle and determining its location relative to other fetal structures. The genital tubercle is the precursor to the penis in the male and precursor to the clitoris in the female. Around 55 days of gestation, the fetus migrates toward the umbilical chord in the male, and toward the anus in the female and at 60 to 70 days it is possible to predict the foal's sex. Knowing the sex of a foal enables a breeder to more accurately appraise the foal's value. Frequently an owner will race all of one sex or the other and sell all of the other sex. Knowing what a mare carries can affect cash flow estimates for the following year.

Daniel and Alec Wildenstein's European champion 2YO and 3YO filly ALLEZ FRANCE (Sea Bird II-Priceless Gem by Hail to Reason) was voted Horse of the Year in France. She won 13 races including the French 1,000 Guineas Gr-1, French Oaks Gr-1, Prix Vermeille Gr-1, Prix d'Ispahan Gr-1, and Prix de l'Arc de Triomphe Gr-1. She traces to La Troienne.

A racehorse breeder must wait one or two generations to be able to discover what strengths and weaknesses are transmitted from a stallion or mare. Only by waiting and inspecting two or three generations of a family's foals can a breeder observe what is present in any given strain. The effectiveness of ATP in a runner's make-up is obviously critical to sustain speed over a distance, but is this a dominant or a recessive factor?

Unfortunately, only pure-breds breed true to type. We are at the mercy of genetic probability factors. "Chance" will play tricks. *Selection for desirable characteristics* enabled great thoroughbred breeders to improve speed and soundness. Trainers have a right to complain about the increasing amount of unsoundness in the breed. They have a right to criticize thoroughbred breeders for tolerating too many structural or physical weaknesses, and not paying enough attention to temperament. Although rate of improvement of speed may have slowed down in the past fifty years there is always the possibility another St. Simon or another Nearco might appear. Most horse farms raise a good yearling. Scientifically-tested feeds are used to ensure the right calcium/phosphate ratio and trace elements and amino-acid additives are supplemented. Of course, I much prefer organically grown feed to chemically treated feed but it is slightly more expensive.

We can manipulate the genes and select for desirable characteristics if we want to make progress. We should avoid lines which transmit weakness. We can aim to improve fertility and soundness simultaneously with the ultimate goal of upgrading speed. These things are achievable. In the end, given a suitable environment, it is the foal's genotype which allows it to make the top 5% of each foal crop. Therefore, we must tinker with thoroughbred genes, and tussle with mother nature.

ALIBHAI (Hyperion-Teresina by Tracery) Leading sire in North America and broodmare sire of Graustark and His Majesty.

The General Stud Book, Volume 1, published by James Weatherby 1808.
The Chemistry of Life (third edition), by Dr. Steven Rose, published by Penguin books 1991.
Arabian Exodus, by Margaret Greely, published by J.A. Allen 1985
Thoroughbred Racing Stock, by Lady Wentworth (rare book) UK.
The Authentic Arabian Horse, by Lady Wentworth (rare book) UK.
Fundamental Genetics, by Dr. Peter J. Russell, published by Harper Collins, 1994.
The Horse, Through Fifty Centuries of Civilization, by Anthony Dent, published by Phaidon Press 1974.
The Blood-Horse Magazine, Lexington, Kentucky.
The Thoroughbred Times, Lexington, Kentucky.
Breeding Racehorses From Cluster Mares, by Dennis Craig, published by J.A. Allen, 1964.
Thoroughbred Pedigree Charts, H.E.Keylock, published by The British Bloodstock Agency, 1935.
Famous Running Horses, by Colonel John F. Wall. Published 1949.
Pedigrees of Leading Winners 1912-1959, by Franklin E. Birch, published by The Thoroughbred Breeders' Association, London, 1960.

Sires and Dams of Stakes Winners, published by The Thoroughbred Owners and Breeders Association & The Jockey Club Statistical Bureau, Lexington, Kentucky, 1979.
Racing and Breeding notes, penned by various English breeders in the period 1650-1750 which are in the possession of Mr. R.E. & G.B. Way, Burrough Green, Newmarket, Suffolk CB8 9NA, England.

Various Stallion registers published in England, France, USA, Australia, Argentina.
Keiba Shushin, Japanese weekly breeding magazine, published by Mr. Aoki, Tokyo, Japan.
The Stallion Book, a Weatherby Publication - various years. UK.
Scientific American, a monthly science journal - various issues. USA.
The Daily Racing Form - various issues. USA.
Various manuscripts of unknown authors. UK.
Pedigrees of Important Female Lines, private handwritten charts by Ken McLean.
Specifications for Speed in the Racehorse - The Airflow Factors, by Dr. W. Robert Cook, published by The Russell Meerdink Company, Menasha, WI, 1993.
American Produce Record, published by Bloodstock Research, Lexington, Kentucky.
The Bloodstock Breeders' Review, published by the British Bloodstock Agency, UK. 1920-1950.
The Great Stallion Book, by Richard Ulbrich, published by Libra Books Pty Ltd, 57 Elizabeth St, Hobart 7001 Tasmania. Australia. 1986.
The Horsetrader, by Patrick Robinson with Nick Robinson, published by Harper Collins, Hammersmith, London. 1993.
Molecular & Cellular Biology, by Wolfe. 1993 edition.

INDEX OF PEDIGREES

Abbot's Trace.. 153
Ajax.. 85
Arazi.. 189
Ayrshire.. 144
Blushing Groom.. 170
Bold Promise.. 149
Chaucer.. 196
Childwick.. 150
Danehill.. 149
Danseur Fabuleux.. 189
Eleanor.. 50
Fair Play.. 176
Felstead.. 147
Flying Fox.. 142
Galopin.. 143
Glencoe.. 81
Gohanna.. 48
Grand Parade.. 147
Gulf Stream.. 196
Hyperion.. 68
Ksar.. 85
Lemonora.. 153
Mahubah.. 176
Man O'War.. 176
Manna.. 147
Mid-day Sun.. 151
Miesque.. 192
Mr. Prospector.. 178
Mtoto.. 132
Nasrullah.. 197
Nearco.. 68
Nearctic.. 151
Northern Taste.. 154
Nureyev.. 192
Partner.. 48
Pasadoble.. 192
Phalaris.. 175
Pondoland.. 148
Raeburn.. 151
Rataplan.. 80
Red God.. 170
Rightaway.. 152
Runaway Bride.. 170
Sanda.. 145
Sans Souici II 155
Sarazen.. 145
Selene.. 196
Serenissima.. 196
Sister Sarah.. 152
Southern Halo.. 182
St. Angela.. 143
St. Josephine.. 148
St. Simon.. 143
Stockwell.. 80
Sunday Silence.. 182
The Scout II 151
Tide-way.. 196
Tracery.. 85
Tourzima.. 199
Wellingtonia.. 83

MR. PROSPECTOR shown winning the Whirlaway Handicap at Garden State Park in smart time. He is one of the most influential sires of the century.

INDEX

A

A.P.Indy 100
Abbot's Anne 152, 153
Abbot's Speed 153
Abbot's Trace 13, 67, 104, 127, 151,
Abdos 149, 190
Abdulla, Hussain Lootah 184
Abdulla, Nasser Lootah 184
Abdullah, Prince Khalid 149, 204
Abercrombie, Josephine 136, 204
Abernant 61,95
Academy Award 88, 100
Acatenango 106
Ack Ack 14, 93, 104
Acton, Robert 186
Ada 44
Adam 85
Adargatis 190, 202
Adeline 143
Adjudicating 89
Admiral's Voyage 89, 109
Admiral Drake 32, 54, 74, 173
Admiring 199
Aethelstan 87, 109
Affirmed 88, 102, 145, 164
Afleet 58, 89, 102
Aga Khan, H.H. 40, 41, 43, 54, 115, 128, 129, 168, 171, 197, 204
Aga Khan, H.H. Karim 168, 169, 173, 204
Again 106
Agrippine 111
Ahonoora 61, 107
Aimee 46, 56, 83, 170, 172, 189
Air Forbes Won 100
Airlie Stud 204
Ajax 45, 68, 83, 85, 86, 87, 109, 151,
Akarad 105, 190, 199
Akaster Turk 15, 25, 50
Akiyama, Yochitaka 71
Akiyda 199
Al Bahathri 169, 170, 171, 172, 188
Al Hattab 105, 179
Al Mamoon 103
Alara 132
Albanilla 199
Alchimist 106
Alcide 105
Alcock's Arabian 15, 25
Aleppo 13, 22
Alexander 81
Alexander, Helen C. 138, 204
Alibhai 13, 67, 83, 85, 95, 192, 193
Alice 82, 85, 203
Aline 152
Aliysa 200
Alizier 110
All Moonshine 32, 54, 121, 196
Alleged 108
Allez France 199, 208
Almahmoud 46, 56, 58, 73, 134, 150, 154, 158, 176, 182, 183, 189, 191
Almanzoar 13
Alsab 14, 104
Altesse Royale 163
Alwah 95
Alwasmi 58
Always Mint 43, 183
Aly Khan, Prince 172, 159, 197
Alycidon 67, 105, 131, 132, 133, 201
Alydar 73, 88, 93, 102
Alydeed 101
Alysheba 88, 102
Amazer 132
Amazon 81, 127, 132
Ambehaving 107
Ambiopoise 107
Ambiorix 55, 107
Ambrose 79, 83
Americus 27, 144
Americus Girl 197
Amie 203
Amphion 110, 111, 141, 151
Amphora 111, 151
Anchora 68
Andover 79
Androcles 107, 113
Angelica 11, 13, 45, 54, 68, 69, 83, 85, 142, 145, 148, 152, 156, 201
Angelus 131
Annette 83
Another Treat 191
Antespend 86
Anthony, Ed 138
Apalachee 55, 108
Apollonia 190
Appiani 106
Apremont 79
Arabian mare 24
Aragon 95
Arathusa 81
Araucaria 76, 83, 85, 141, 152, 154
Arazi 55, 61, 98, 131, 157, 169, 180, 188, 189, 209, 210
Arcadia 175, 196
Archduke 24
Arctic Tern 56, 88, 102, 134, 154,
Argentina 85
Arietta 46, 149
Aristophanes 67, 95, 192
Arjaman 106
Armageddon 104
Armos 100
Arrowfield Stud 204, 205
Art Paper 86
Artichoke 100
Arts and Letters 108, 179
Askywalker 103
Asmena 163, 190
Aspidistra 46, 56, 183
Assault 145
Asta 85
Asterus 13, 45, 83, 109, 130, 142,
Astor, Lord 198
Astor, Sir John 204
Astronomie 46, 190
At the Half 179
Atabor 106
Atalanta 142, 144, 166, 167
Atan 88, 102
Athanasius 105
Athasi 128
Atlantic 110
Auditing 110
Aurelius 164
Aureole 55, 67, 95
Auricula 79
Aurora 132, 133
Auspicante 86
Australian 103, 113
Automation 79
Auxey 86
Awaasif 188
Ayacanoora 79, 88, 85, 152, 154
Ayr Laddie 79
Ayrshire 142, 144, 171

B

Baalim 106
Bacchante 50, 80, 81
Bachelor's Double 132
Bacteriophage 110
Badger Land 207
Badruddin 43, 51, 105, 121, 129
Baederwood 103
Bahram 129, 130 132
Bahram 40, 64, 67, 88, 105, 127, 129, 130, 132
Bahri 98
Baillamont 169
Balanchine 188
Balconaje 107
Bald Galloway 14, 23 50
Balding, Ian 173
Balidar 95
Balior 100
Ballade 47
Ballyheron 128
Ballymacoll Stud 204
Ballymoss 164
Balmerino (NZ) 207
Banshee 199
Banstead Manor Farm 204
Banter 23, 44, 53, 83
Baramul Stud 205
Barathea 101, 205
Barbara Burrini 202
Barbizon 26
Barcaldine 24, 103, 113, 141
Barcarolle 26
Barnett, W & R 128, 203
Barry, Clifford 136
Bartlett's Childers 13, 19, 24
Basajaun 104
Bashaw 114
Basse Terre 85
Bassett's Oxford Bl.Sh.Arab 15, 24
Bassett Arabian 15, 24
Bastro 22
Bates Motel 97
Batonnier 108
Battle Joined 104
Bay Bolton 24, 25, 49
Bay Middleton 44, 113, 143
Bay Richmond 114
Bay Ronald 68, 87, 94, 95, 127
Bayardo 54, 68, 94, 95, 196
Bayraan 168
Be My Guest 90, 101
Beachy Head 79
Beam 163
Beau Cheval 106
Beau Gar 110
Beau Genius 100
Beau Pere 56, 106
Beaulivre 106
Beaune 154
Beck, Antony 207
Beck, Graham 186, 204, 207
Beeswing 44, 53, 144, 166
Belfonds 79

Belgrade Turk 14, 15, 24
Believe It 103
Bella 155
Bella Paola 200
Belladrum 79
Belle of Troy 202,
Bellini 108
Belmont, August 202
Ben Battle 79
Ben Brush 26, 104
Bend Or 13, 14, 80, 96, 112, 145, 146, 155, 166, 167, 176, 202
Beningborough 81
Benson, Martin 129
Bering 88, 102, 154, 160
Bertrando 103
Bess Brimmer Mare 26
Best Turn 88, 97
Bethel's Arabian 15, 19, 50
Better Boy 107
Better Loosen Up 207
Better Self 14, 46, 56, 203
Betty Leedes 18
Bey 109
Big Game 88, 105
Big Spruce 106
Bijou 85
Bikala 145
Bimelech 14, 56, 202
Binalong 103
Birdcatcher 13, 44, 112, 143, 161
Birdcatcher 166
Biribi 108
Birkhahn 106
Biscay 47, 58, 95
Black Cherry 127, 198
Black Domino 171
Black Duchess 45, 54, 68, 129, 135
Black Hearty 25
Black Morocco Barb 17
Black Polly 55
Black Ray 45, 171
Black Servant 104, 202
Black Tie Affair 102, 179
Black Toney 104, 112
Blacklegs 22
Blacklock 12, 29, 44, 53, 80, 112,
Blade 100
Blair Athol 79, 162, 166
Blaise 148
Blakeney 107
Blanche 127, 178
Blandford 12, 13, 14, 31, 45, 51, 54, 67, 88, 105, 112, 127 to 135, 145, 148, 166, 170, 175, 176
Blank 11, 44, 48
Blaze 49, 50
Blazonry 128
Blenheim 13, 40, 51, 67, 105, 127, 128, 129, 132, 134, 135, 162, 211
Bletchingly 58, 95
Blink Bonny 155, 166
Blinkhoolie 79, 152, 153
Bloodroot 46
Blue Bear 128
Blue Ensign 108
Blue Green 148
Blue Larkspur 14, 45, 104, 176, 178,
Blue Peter 13, 96
Blue Swords 104, 182
Bluebird 101, 205
Bluebook 31
Bluette 85
Blush with Pride 169
Blushing Groom 47, 56, 58, 88, 90, 93, 98, 159, 167 to 173, 180 to 189
Blushing John 169
Boarding School Miss 77
Bob's Return 164
Bob Booty 80, 161
Boiard 155
Bois Roussel 54, 74, 108, 121, 130, 160, 162
Bold Bidder 100, 167, 181
Bold Commander 100, 179
Bold Forbes 100
Bold Irish 46
Bold Lad (Ire) 100, 181
Bold Lad (USA) 100
Bold Promise 149
Bold Reason 154
Bold Ruckus 100
Bold Ruler 46, 67, 72, 98, 100, 149, 168, 182, 203, 211
Boldnesian 100
Bolero 72
Bolero Rose 72
Bolton 26
Bona Vista 13, 80, 96
Bonny Black 10, 12, 25
Boojum 104
Boola Brook 46, 170, 171, 189
Border Run 73
Bosra Sham 178
Boston 12, 141
Bosworth 106, 128, 172
Bott, Tony 205
Boudoir II 46
Boussac, Marcel 85, 190, 198, 202
Boutin, Francois 183, 190, 191, 192
Boutin, Madame Lucy 192
Bowles, Paddy 189
Bozzetto 174
Brabham 26
Bradley, Colonel E.R. 199, 202
Brady, John 189
Bramble 104
Brantome 105, 130, 131, 135, 165
Breadalbane 155
Briar-root 146
Briartic 101
Bridaine 129
Bridge of Allan 150, 175
Brief Truce 98, 205
Brimmer 22, 24
Brimmer mare, The 24
Brinsfield, Les 139
Brisk 13
Bristol's Hog 24
British Empire 96
Broad Brush 14, 98, 104
Brocco 61, 97, 157, 205
Bromus 65, 147
Broomstick 12, 45, 54, 104
Brown Bess 44, 171
Bruce, Benjamin & Sanders 11
Bruleur 85, 107, 113, 199
Bubbling Over 164
Buchan 110
Buckaroo 96
Buckfinder 96
Buckpasser 47, 96, 149, 157, 158
Buckpasser 176, 178, 179, 191, 199
Bucksplasher 31, 47, 56, 96
Buckwheat 147
Budgett R..A. 204
Bull Dog 13, 32, 45, 54, 57, 74, 83, 86, 87, 109, 121, 142, 145, 149, 170, 176, 178, 179, 211
Bull Lea 87, 94, 109, 182, 203
Bull Page 109, 127, 203
Bulle Rock 13
Bunting 26
Bunty Lawless 203
Burgh, Major J.H. 204
Burton's Natural Barb 21
Busanda 178
Busher 199, 202
Businesslike 178, 202
Bustard 22
Busted 105, 131, 132
Bustino 105, 164, 188
Bustler 15, 18, 19, 22, 23, 24, 25, 26
Bustler Mare, The 25
Butters, Franck 124
Buzz My Bell 43, 194
Buzzard 12, 81, 110, 113
Byerley Turk, The 10, 15, 16, 17, 19, 21, 22, 23, 25, 48, 87, 113, 141

C

C'Mon Let's Dance 73
Cachuca 79
Cacoethes 88
Cade 11, 49, 50, 113
Cadeaux Genereux 188
Cadeaux Genereux 95
Caerleon (by Phalaris) 174
Caerleon 13, 58, 101
Caerleon 174., 180, 188
Cahill Road 51, 102
Cain 143
Calash 50, 81
Caller I. D. 61, 101, 157
Calumet Farm 204
Cambaules 77
Cambridge Stud 205
Cambyse 113
Cambyses 107
Camel 76, 77, 83, 112
Camelia 79
Cameronian 198
Campanula 128
Campanule 85
Candy Spots 179
Canip 68, 69, 85, 88
Cannon, Jim 136
Cannonade 100
Canny Lad 205
Canny Lad 95
Cantelo 164
Canterbury Pilgrim 45, 54, 68, 69, 87, 127, 132, 148, 151, 163, 170, 174, 175, 193, 195 to 197
Cap and Bells 163
Cap d'Or 88
Capo di Monte 198
Capote 100
Captain Byerley 12
Captain Cuttle 103
Captain Cuttle 128
Cara Rafaela 149
Caravelle 190

Carbine, 11, 68, 69, 147, 166, 171, 201
Cardanil II 107
Careless 19
Carlisle's Turk 23
Carnegie 165
Caro 47, 57, 98, 167
Carpet Slipper 129
Carpet Slipper 175
Carr, Ted 136, 188
Carrozza 198
Castrel 12, 44, 110, 112
Catnip 151, 197, 201
Catrail 101, 158, 205
Cavaliere d'Arpino 108
Caveat 100
Cellini 181
Cequillo 46
Cestus 79
Challenger II 111
Chamant 79, 167
Chambure, Comte R. de 204
Chandler, Alice 136, 204
Chandler, Dr. John 136
Chanoinesse 147, 196
Chapel of Dreams 73
Charles O'Malley 128, 129, 135
Charlotte West 77
Chatanooga 79, 83, 141, 150
Chatelaine 163
Chatham 79
Chatherton 176
Chaucer 12, 32, 45, 52, 54, 67, 68, 69, 87, 108, 127, 166, 170, 174, 175, 195, 196, 209
Cheery 68
Chelandry 45
Chenery, C.T. 203
Cherfas, Jeremy 36
Cherokee Colony 108
Cherokee Run 98
Cherry Lass 163, 198
Cheveley Park Stud 204
Chief's Crown 31, 86, 89, 101, 205
Chieftain 100
Chillaby Turk 15
Childers 112
Childers mare 22
Childwick 150, 166
Chit Chat 176
Chop Chop 109, 154, 203
Chouberski 85, 107, 113
Chulmleigh 164
Churrinche 104
Ciboulette 47
Cicero 142, 162
Cigar 111, 145, 151, 161, 183, 194
Cinderella 176
Cinna 45
Cipayo 105
Circus Plume 163
Citation 47, 64, 164, 203
Clamart 79, 83, 85, 203
Clare 77, 80, 81, 83
Clarion III 107
Clarissimus 130, 131, 176
Claro 111
Classy'n Smart 58
Clay, Robert 136
Clever Trick 13, 101
Clover 79
Clubfoot 25
Cobweb 77
Cohoes 105, 134
Colin 104
Collingrove Stud 204, 205
Collins, Joss 3, 89, 136
Colombo 13, 96, 129
Colonel Collins 101
Colonel St. Leger 10
Colonial Affair 108
Colorado 13, 174
Colorspin 163
Commander in Chief 57, 163
Commando 104, 145
Commemorate 102
Comrade 165
Comtesse de Loir 192
Comus 113
Con Brio 108
Concern 104
Concertina 32, 52, 129
Concerto 61
Conductor 113
Congreve 13, 85, 104
Consols 108
Cooke, Jack Kent 204
Coombs, Leslie 204
Cooper, Alan 136, 192
Cooper, Tom 57, 192
Coppin mare 22, 24
Copyright 13, 45, 85, 104
Coquistador Cielo 89, 102
Corindon 104
Cormorant 93
Cornish Prince 191
Coronach 67, 88, 103, 129, 164, 167
Coronation V, 165, 190, 199
Corrida 165, 190
Cosmah 182
Cosquilla 86, 149, 198
Cothram, Dr. Gus 71, 72, 75, 76
Count Amber 110, 178
Count Fleet 86, 110, 145, 178
Count of Honor 110
Count Turf 110
Counterpoint 110
Coup de Folie 47, 58, 134, 150, 190, 191
Coup de Genie 58, 89, 134, 191
Court Martial 13, 46, 54, 62, 96, 173, 198
Courtly Dee 26, 47
Couturie, Madame J. 204
Cox's Ridge 88, 97, 150, 179
Cozzene 97
Crafty Admiral 109, 145
Crafty Prospector 61, 102
Cranach 180
Cream Cheeks 18, 49, 50, 51
Crepello 46, 105, 112, 131, 132, 133 135, 162, 201
Crepuscule 131, 132
Cressida 51
Cribb, Ross 149
Crick, Dr. Francis 71, 119, 120
Criminal Type 88
Crimson Saint 72, 73
Crimson Satan 72
Crisford, Simon 188
Crozier 107
Crystal Glitters 169
Cub Mare, The 44, 114
Cuddles 43
Cullen Arabian, The 10, 14
Culver, Francis 27
Cunningham, Prof. E.P. 19, 20, 62
Cure the Blues 97
Curwen's Arabian Mare 17, 22
Curwen's Bay Barb 15, 16, 17, 18,
Cutaway 46
Cutty Sark 21, 26
Cyane 97, 178
Cybiane 131
Cyllene 13, 45, 68, 80, 96, 148, 153,
Cypron 49, 50

D

D'Arcy, James 17, 18, 21, 141
Daffodil's Dam 23
Daffodil 23
Dahar 101
Dahlia 205
Daingerfield, Major 201
Dalham Hall Stud 189, 191
Dalmary 129, 211
Dalry 107
Dalsaan 97
Damascus 47, 57, 69, 87, 88, 109, 178, 211
Dame Masham 176
Dan Cupid 102
Dance Smartly 58, 89
Dancing Brave 64
Dancing Count 178
Dancing Moss 100
Danehill 31, 47, 58, 61, 89, 90, 101, 149, 154, 158, 180, 205, 212
Danewin 212
Danseur Fabuleux 189
Dante 13, 67, 79, 87, 100, 115, 133
Danzero 205
Danzig 13, 31, 47, 52, 58, 61, 89, 90, 101, 145, 149, 180, 212
Darby Dan Farm 183, 181, 204
Darcy's Black Legged Royal mare 22
Darcy's Chestnut Arabian 14, 18
Darcy's Layton Barb 24
Darcy's Old Montagu 22, 25
Darcy's Old Sorrill 22
Darcy's Royal Montagu Mare 22
Darcy's White Turk 14 to 19, 21, 22,
Darcy Yellow Turk, 14, 18, 23, 48
Darcy;s Layton Vioet Grey 21, 22
Darius 100
Dark Ronald 106
Dark Star 164
Darley Arabian, The 10, 12 to 19, 23 to 25, 50, 87, 112, 141
Darling, Fred 130
Darshaan 98, 199, 200, 211
Dart 13
Darwin, Charles 63
Dastur 129
Davill's Old Woodcock 23
Dawson Dick 128,129
Dayjur 89, 101, 186, 188
Decidedly 164
Decker, Rich 136
Dehere 61, 157
Deiri 87, 109
Delahooke, James 136
DeLancey, James 114

Delleana 128, 132
Delsy 47
Demons Begone 108, 157
Demure 149
Deputed Testamony 95
Deputy Minister 93, 101
Derby, Lord 67, 68,127, 159, 170, 195
Derring-Do 100
Derry Meeting Farm 204
Desmond 12, 147, 152, 175, 176
Determine 67
Detroit 165
Deux Pour Cent 87, 109
Devil's Bag 57, 88, 97, 149
Dewar, Lord 198
Diamond Jubilee 52,163
Dick Andrews 81
Dick Burton's Barb 25
Dickens, Charles 71
Dicky Pierson 21
Dido 114
Diesis 102,188
Digamist 169
Dike 106
Diminuendo 115, 188
Dinah Mare 27
Dinmont 26
Dinner Partner 189
Diomed 12, 50, 81
Discovery 11, 46, 113,203
Dismal 11
Dispersal 97
Display 11, 113
Disraeli 142
Distant Relative 97
Distant Shore 148, 175
Distant View 89
Diversion 44, 144
Dixie Brass 101
Dixieland Band 13, 90, 101
Djebel 27, 46, 55, 67, 83, 107, 113 150, 153, 165, 190, 199
Djeddah 107
Djezima 46
Dobbins 26
Dodsworth's Dam 25
Dodsworth's Turk 14
Dodsworth 21, 25
Dollar 85, 107, 113
Dolly Varden 77
Dolphin Street 101
Dominion 100
Domino 12, 14, 45, 54, 104, 145, 167
Don't Say Halo 97, 205
Don Juan 83
Dona Roca 176
Donatello II 14, 67, 105, 128, 131, 132, 133, 156, 166, 176, 201
Doncaster 13, 79, 80, 85, 96, 112, 132, 133, 156, 166, 176, 201, 144
Dongola 68, 196
Donovan 142
Doremi 202
Doricles 107
Double Life 132
Doubles Partner 189
Douranee 149
Doyoun 98
Dr. Blum 104
Dr. Devious 107
Dr. Fager 104, 149
Drap d'Or 128
Drift 196
Driver 81
Dubasoff 173
Duc d'Orleons 27
Duke of Chartre's Hawker 24
Duke of Westminster 142
Dunlin 176
Dupre, Francois 173, 200, 204
Durban 199
Durbar II 26, 27, 162, 199
Dust Commander 100
Dynaformer 97

E

Eagle Eyed 101, 149, 154
Earl of Godolphin, 11
Earl of Rosebery 198
Eastern Echo 87
Easton 129
Easy Goer 88, 93
Ebony 22
Echidna 80
Eclair 170, 171, 182, 173
Eclat 171
Eclipse 10 -13, 22, 25, 29, 44, 50-53, 75, 76, 95, 104, 108-112, 166
Economist 80
Efisio 95
Eight Carat 47
Eight Thirty 13, 104
Eillo 89
El Gran Senor 90, 101
Eleanor 50, 51, 80, 81
Ellen Horne 44
Ellesworth, Rex 171
Elmendorf Farm 203
Elocutionist 157
Emali 170, 171, 172, 189
Embrujo 104
Emilius 51
Emirates Park Stud 205, 212
Emperor of Norfolk 27
Endeavour II 96
Enquirer 26
Eohippus 35
Epervier Bleu 149
Equadif 149, 150
Equal Venture 192
Equestrian 26
Equipoise 104
Equus 35, 36
Erhaab 89
Eryholme 68
Espresso 179
Eternal 104
Ethnarch 170
Eve Stud Limited 204
Evelina 81
Everlasting 146
Exbury 104, 165
Exceller 95
Exclusive Era 102
Exclusive Native 102
Exit to Nowhere 58, 98, 134, 191
Explodent 101
Ezzoud 101, 188

F

Fabuleux Jane 189
Fair Harvest 111
Fair Isle 175
Fair Play 103, 113, 175, 176, 202
Fair Trial 12, 18, 46, 54, 62, 96, 112, 129, 166. 173, 198
Fairway 13, 96, 111, 129, 174, 195, 196
Fairy Gold 45, 176
Fall Aspen 47
Falstaff 101
Fancy Ribbons 43
Fappiano 32, 51, 58, 61, 89, 102, 149, 183, 211
Far North 90, 101
Faraway Son 107
Fares Farm 204
Farish, Will S. III, 136, 204
Farma Way 100
Faronia 144, 155
Fastnet 174
Favourite 23
Fearnought 11
Fearnought mare 26, 27
Felicitation 130
Felkington 147
Felstead 147
Fenwick Barb 14, 15, 23
Feola 46, 55
Ferro 105
Festa 45, 166
fetal-sexing 215
Field, Derek 136
Fiesta Lady 31
Fifinella 130
Fighting Fox 109, 212
Figure 114
Filagree 77
Firdaussi 13, 149
Fire the Groom 169
Fit to Fight 100, 103
Flagette 149
Flamboyant 13, 85, 104
Flaming Page 94, 127
Flaming Top 46, 166
Flaming Vixen 111
Flanders 179, 209
Flares 109
Flattery, Jacinta 118
Fleet Nasrullah 98
Flight 80
Flirting Around 108
Flora 24
Floranthe 80
Florizel 12
Florizel II 50, 108
Flower Bowl 149
Fly So Free 87, 109
Flying Childers 13, 18, 19, 22, 25
Flying Duchess 143, 144
Flying Fox 13, 45, 68, 83, 85, 109, 111, 141, 142, 151, 166
Flying Paster 100
Flying Spur 212
Flying Vixen 151
Flying Water 169
Folle Rouse 168
Forli 67, 95, 192
Formidable 95
Forrester, Frank 27
Fortino II 98
Forty Niner 58, 61, 92, 93, 102
Foveros 207

Fox 49
Foxbridge 106
Foxhunter 106, 111
Foxlaw 106
Frampton, Tregonwell Mr. 21, 141
Fraxinella 81, 146
Free For All 104
Fregin, Dr. G. Frederick 76
Fresh Air 95
Fresnay-Le-Buffard 190, 193
Friar's Daughter 129
Friar Marcus 129
Friar Rock 13
Frizette 45, 190, 199
Frosty the Snowman 108, 133
Full Pocket 103
Furness, Lord 150
Future Storm 43, 101

G

Gabriella 27
Gadabout 80
Gaga 178
Gaines, John 186, 204
Gainsborough 12, 14, 67, 68, 94, 95, 151, 170, 172, 196
Galcador 162, 190
Galicia 68, 142, 148, 153
Gallant Fox 27, 109, 212
Gallant Man 108, 112, 145
Gallant Romeo 108
Galliard 142
Gallinule 148, 153, 167
Galloping Queen 26
Galopin 12, 40, 45, 52, 68, 85, 108, 112, 141-144, 147-153, 176, 196
Galton, Francis 63
Game Dame 149
Gana Facil 51, 61, 183
Gardefeu 85, 107, 113
Gardenia 79
Garron Lass 127
Gas 142
Gascoigne's Foreign Horse 23
Gato del Sol 26
Gay Crusader 48, 88, 153
Gay Gallanta 178
Gay Gamp 153
Gay Lussac 179
Gay Mecene 190
Geiger Counter 102
General Assembley 88, 100
Generous 163
Gentry, Tom 204
Gevney 86
Giantess 44, 50, 81
Gilded Time 61, 87, 109, 157, 211
Gimcrack 24
Glad Rags II 47
Gladiateur 27
Glave 148
Glencoe 12, 27, 31, 44, 50, 53, 54,
Glorious Song 57, 149, 183
Go and Go 101
Go For Gin 58, 165
Go Marching 108
Godolphin Arabian, The 10, 11, 14-16, 19, 26, 49, 50, 87, 113
Gohanna 24, 49- 51, 81, 141, 146
Gold Beauty 89
Gold Bridge 60, 211
Gold Digger 58, 88, 178
Gold River 173
Gold Splash 169
Gold Stage 102
Gold Sun 111
Golden Attraction 43
Golden Fleece 127, 163
Golden Locks 24
Golden Thatch 207
Gondolette 45, 68,145, 150, 170, 172, 195, 196, 197
Gone West 57, 58, 61, 89, 93, 102
Good Goods 104
Good Manners 98
Goodbody, Michael 189
Gower Stallion 22
Grand Geraldine 147
Grand Lodge 86, 205
Grand Marnier 147
Grand Parade 13, 147
Grand Plaisir 200
Grasshopper 19, 24, 25
Grasshopper mare 25
Graustark 47, 57, 94, 108, 191, 192
Great Above 104
Grecian Banner 47, 58
Green Dancer 101, 180
Green Desert 26, 89, 101, 188
Green God 168
Green Means Go 180
Gresley's Arabian 25
Grey Classic 58
Grey Dawn II 167
Grey Flight 46, 56, 134
Grey Royal 23\
Grey Sovereign 61, 62, 98
Grey Whynot 23
Grey Wilkes 49
Greyhound 19
Gribbon, John 36
Griffin, Thomas 127
Grindstone 43, 61, 165, 183, 194
Groom Dancer 58, 98, 169
Grosveneor 88, 97
Guard's Tie 110
Guggenheim, Cpt. Harry 208
Guiccioli 80, 83, 161, 160
Gulch 58, 89, 102
Gulf Pearl 105
Gulf Stream 95, 196
Gummo 98
Gundomar 106
Gunther, Herr 167

H

Habitat 56. 62. 86, 97
Hachazo 104
Hail To Reason 57, 88, 93, 97, 179 182
Hale's Turk 23
Halifax, Lord 204
Hall Walker, William 127, 197, 201
Halo 47, 56, 57, 61, 88, 90, 93, 97, 134, 154, 156, 167, 176, 182, 183, 190, 191, 205
Hambletonian 112
Hammerkop 147
Hampton 14, 45, 54, 68, 69, 94, 95, 112, 142, 144, 148
Hancock A.B. "Bull" Hancock II 204
Hancock A.B. Snr 203
Hancock, Arthur B. III 136, 204
Hancock, Seth 92, 136, 204
Hannah 79
Hanover 12
Hansel, 102, 178, 186
Happy Landing 129
Harbor View Farm 204
Haricot 144
Harkaway 79, 83, 143
Harpalaice 80, 81, 146
Harpy 111, 151
Harry of Hereford 178
Hastings 103, 111, 113, 176
Hasty Reply 169
Hatchet Man 105
Haun, Marianna 71, 72, 75, 76
Hautboy 16, 20, 22, 25, 49
Havresac II 12, 54, 67, 68, 85, 86, 88, 108, 145, 151, 197
Hawkster 98
Hayata, Dr. Koichiro 71, 186
Hayes, Colin S. 205
Head, Alec 154, 204
Headley, Hal Price 203
Heavenly Prize 179
Hector Protector 178
Height of Fashion 167, 188, 198
Heldifann 199
Helene de Troie 202, 203
Helicon 202, 203
Hellenic 200
Helmsley Turk 14, 15, 18, 19, 23, 25
Herbager 67, 106, 149
Hermit 14, 29, 45, 54, 79, 147, 148, 153, 155, 162, 175, 176, 196
Hernan, Michael 136
Hernando 191
Herod (King Herod) 12, 14, 29, 44, 49-54, 77, 81, 107, 110, 113, 166
Heroine of Lucknow 79
Herold 106
Hethersett 107, 113
Heume 146, 155
High Hat 95
High Time 145
High Top 100
High Voltage 46
Highclere 198
Highflyer 12, 23, 44, 50, 81
Highland Ladie 25
Highland Park 102
Highlight 198
Himyar 14, 104
Hindoo 12
Hinton, Monty 139
Hippia 79
His Grace 129
His Majesty 47, 57, 93, 108, 133 149, 211, 212
Hobby Mare, The 24
Hobday, Sir Frederic 167
Hobgoblin 24, 26
Hogan, Patrick 90,205
Hogarth 105
Hoist the Flag 47, 57, 108, 190
Hold Your Peace 108, 205
Holder, Dr. Richard 215
Holderness Turk 15, 23
Holliday, Major L.B. 204
Hollingsworth, S.D. 151
Holy Bull 10, 61, 104, 208

Honeycomb Punch 25
Honeysuckle 54, 144
Honeywood's Arabian 15
Honiton 152
Honora 153
Hooper, F.W. 204
Hopespringseternal 176
Hornsea 143
Hors Concours 68
Hot Grove 169
Housebuster 98
Hugh Lupus 107, 167
Humming Bird 77
Humphrey Clinker 113, 167
Hunt, Nelson Bunker 204
Hunter's Moon 32, 54, 196
Huntercombe 100
Hurry Harriet 131
Hurry On 11, 67, 88, 103, 111, 113, 128, 166
Hutton's Grey Barb 22
Hutton's Spot 23
Hydro Calido 58, 150, 191
Hypericum 198
Hyperion 12, 14, 31, 32, 46, 52, 54, 67-69, 94, 112, 121, 132, 144, 151, 153, 159, 162, 164, 166, 170-172, 176, 181, 192-198, 203

I

Icecapade 13, 93, 101
Idas 77
Idomeneo 149
Ifrad 169
Ile de Bourbon 127
Imagery 150
Imbrolglio 106
Impromptu 61
In Reality 61, 93, 103, 113
In The Wings 101
inbreeding 51
Inca Queen 26
Indian Chief 107
Indian Hemp 98
Indian Ridge 107
Indian Skimmer 188
Indiana 77. 85
Industry 111
Ingham, Bob & Jack 205
Intent 103
Intentionally 103, 113
Intermezzo 164
Intrepidity 163
Intriguing 47
Inverness Drive 107
Ion 143
Irish Ball 145
Irish Bird 145
Irish Castle 100
Irish Lad 27, 199
Irish Lass 144
Irish River 47, 98, 191
Irish Tower 100
Iron Reward 56
Irony 85, 142
Irwin, Lord 204
Isabel 68
Isinglass 105, 163, 166
Island Whirl 96
Isoletta 68
Isonomy 45, 54, 105, 112, 148, 166
It's in the Air 89

J

J. O. Tobin 168, 198
Jacinth 168
Jacinto 100
Jacobs, Ethel D. 204
Jacopo 171
Jade Hunter 102
Jade Robbery 89, 191
Janus 11, 27
Janus Mare, The 26
Jardin, Sir Robert
Java Gold 108
Jefferson, Colonel James 77
Jeremy Diddler 77
Jester 96
Jet Pilot 128
Jet Ski Lady 163
Jig Time 102
Jigg 12, 48, 49, 50, 51, 113
Joe Andrews 81
Joel, H.J. 204
Joel, S.B. 148
Johann Quatz 101, 191, 205
John O'Gaunt 87, 105
John P. Grier 104
Johnstown 178
Jolie's Halo 88, 97
Jones, Ben 209
Jones, Brereton 136
Jones, Johnny 136
Jones, Warner L. Jr. 204
Julia 44, 51
Jungle Queen 111
Jupiter Island 95

K

Kahyasi 127, 180
Kalamoun 98
Kamar 47
Kantar 77
Kaoru Star 95
Kasbah 77, 85
Katia 85
Keene, James 201
Kenmare 98
Kent Cooke, Jack 3, 86
Kerala 57
Kershaw, Allen 186
Kettledrum 144
Key Bridge 46
Key to the Mint 108, 183, 191, 193
Khaled 56, 67, 95, 171
Kilroy, W.S. 204
King Alfonso 26
King Charles II, 17, 18, 141
King Fergus 12, 81, 112
King James 104
King of the Tudors 95
Kaapstad 97
King Pellinore 181
King Salmon 110, 129
King Tom 13, 31, 50, 52, 76, 79, 83,
Kingcraft 79
Kingmambo 57, 58, 61, 89, 102, 160, 192
Kingston Rule 88
Kirwin, Peter 136
Kisber 146
Kite 83
Kizil Kourgan 77, 85, 199
Kizil Sou 77
Klairon 46, 107, 118
Knight of Kars 79
Knight of St. George 79
Knight of St. Patrick 79
Knight of the Garter 77
Known Fact 57, 103, 157, 190
Koch, Gus 136
Kotishaan 199
Kris 102, 154
Kris S. 88, 93, 97
Ksar 12, 45, 54, 67, 77, 83, 85, 107, 113, 131, 160, 165, 199
Kyle's Our Man 103
Kyloe 129

L

L'Enjoleur 96
La Dauphine 141
La Farina 45
La Flambee 151
La Fleche 163, 166
La Papagena 86
La Saggittaire 141
La Troienne 21, 45, 73, 83, 199, 202, 203
Labrador 104
Labus 105
Lac Oumiet 108
Lacerta 143
Lacydon 105
Lady's Secret 31, 208
Lady Americus 144
Lady Angela 52, 67, 86, 151, 154, 182, 192, 193, 203
Lady Carla 163, 188
Lady Cynosure 55, 171, 188
Lady Electra 171
Lady Evelyn 111, 145
Lady Josephine 27, 43, 54, 55, 111, 127, 144, 197, 198
Lady Juror 27, 54, 111, 173, 193, 198
Lady Langden 144
Lady Lumley 153
Lady Mischief 147
Lady of Pedigree 202
Lady Victoria 154, 182
Lalun 154
Lambton, George 43
Lammtarra 101, 127, 180, 188
Landers, Bill 136, 138
Landgraf 105
Landscape 24
Lane's End Farm 190
Langar 83
Laramie Trail 85, 95
Last Tycoon 101
Lavendula II 46, 55, 73, 166, 173
Layton's Grey Barb 21
Le Comte 12
Le Fabuleux 55, 61, 108, 183, 189
Le Haar 105
Le Lavandou 107
Le Levanstell 107
Le Pacha 108
Le Roi Soleil 155
Le Samaritain 110
Le Sancy 110
Lea Lark 182
Leader, Ted 129

Lear Fan 88, 97
Leedes Arabian 14, 15, 18, 19, 23, 50
Lemberg 142, 147, 148, 153
Lemonora 153
Levmoss 107, 165
Lexington 12, 44, 54, 166, 176
Licata 199
Limestone 79
linebreeding 51
Lines of Power 102
Lion Cavern 57, 58, 102
Lister Turk, The 14, 19, 24, 50
Literat 106
Little Fairy 143
Little Hut 46, 56
Little John 23
Live Oak Farm 204
Lively One 97
Local Talent 101
Lomond 90, 101
Lord Avie 97
Lord Clifden 14, 94, 95,144, 145
Lord Derby (see Derby, Lord)
Lord Fairfax's Morocco Barb 14
Lord Fairfax's Sultan Arab. 14, 15
Lord Lyon 79
Lord Ronald 79
Lorenzaccio 107
Louvoirs 105
Love the Groom 169
Love Wisely 68
Loved One 68, 69, 170, 195, 196
Lowe, Bruce 20, 21, 26
Lucky Debonair 105
Lukas, D. Wayne 73, 209
Lunc, Charles 18
Lupe 163
Lure 31, 89, 101
Luskin Star 95, 149
Luthier 107
Lyndhurst Stud 205
Lyphard's Dancer 73
Lyphard 47, 57, 90, 101, 154, 159, 167, 179, 180, 181, 191
Lypheor 101
Lyrism 191
M
Mabee, John & Betty 204
Macaroni 12, 13, 44, 54, 79, 155, 166, 176, 203
Machiavellian 58, 61, 88, 102, 134, 160, 191
Macip 199
Mad Hatter 176
Madcap 176
Maggie B.B. 44, 54
Mah Iran 43
Mah Mahal 43, 121, 128, 145
Mahmoud 10, 13, 40, 43, 46, 54, 55,
Mahogany Hall 178
Mahon, Professor Garry 19, 20
Mahubah 13, 45, 175, 176
Maid Marian 68, 175
Maid of Masham 44
Maid of the Mint 68, 201
Majestic Light 102, 178
Majestic Prince 102, 164
Majesty's Prince 108
Majideh 46
Makeless 25
Maktoum, Sheikh Ahmed 57, 185
Maktoum, Sheikh Hamdan 57, 185, 186, 204
Maktoum, Sheikh Maktoun H. H. 57, 184, 185, 186, 188, 189, 204
Maktoum, Sheikh Mohammed 57, 184, 185, 186, 188, 189, 204
Maktoum, Sheikh Rashid 57, 184
Malva 128, 129
Man O'War 11, 13, 45, 54, 103, 111,
Manila 94, 101, 131
Manna 13, 96, 130, 146, 147
Mannie Grey 45
Marasali 105
March Past 96
Marchetta 132
Marco 103, 113
Marconigram 153
Marcovil 11, 67, 103
Marfa 100
Marguerite 45
Marguerite de Valois 32, 121, 149
Maria's Mon 157
Marju 101
Mark of Esteem 200
Marked Tree 102
Markham Arabian 14, 17
Marmion 80, 81
Maroof 89, 90
Marpessa 77, 80, 81, 83, 143
Marquis de Conflans, The 27
Marscay 205
Marsh, Marcus 129
Marshall, Joe 128
Marske 49, 112
Marsyas 109
Martha Lynn 44
Martial Loan 132
Mary Seaton 148
Mason, P. 127
Massaret 192
Massey's Black Barb 14, 22
Massine 108, 150, 165
Masthead 06
Matahawk 106
Matchem 11, 44, 49, 50, 51, 103, 113, 136
Mathet, Francois 173
Matiara 154
Matrice 96
Matun 109
Maximova 47
Mayer, Louis B. 203
McAnally, Ron 86
McGrath , J. 204
McLean, Pope 204
McLean, Pope Jr. 136
McNamee Sullivan, J. 168
Meade's Celer Mare 26
Meadowlake 108
Medeah 151
Media Starguest 101
Medley Marem The 26
Medora 166
Melbourne 11, 44, 53, 94, 103, 113, 144, 166
Mellon, Paul 204
Memoir 52, 146, 163
Mendel, Gregor 116, 117, 118
Menow 13, 96, 170, 176, 178
Mercedes Won 100
Mercer, Joe 189
Mercury 49, 50, 81
Meridian 164
Merlene 149
Merlin 24, 25
Merlin Mare, The 24
Merope 142, 143. 144
Merry Agnes 148
Merry Gal 148
Merry Token 176
Merry, James 80
Merychippus 35
Mesohippus 35
Messara, John 205
Metfield 100
Meyerhoff, Robert 204
Mi Cielo 102
Miami 166
Mid-day Sun 150, 162
Midstream 129
Miesque 57, 61, 167, 191- 193, 208
Mieuxce 46, 55, 108, 131, 132, 150
Migoli 108, 165
Milford 79
Military Plume 88. 97
Mill Lady 191
Mill Native 102
Mill Reef 47, 57, 86, 98, 115, 163, 165, 173, 178, 179, 181
Miller, Percy 129
Miller, Sir James 146
Mincio 131, 132
Miner's Mark 58, 89, 102
Mineral 145
Minnesota Mac 104
Minoru 32, 68, 128, 196, 197
Minthe 80
Minting 80, 148, 201
Mira Femme 47
Mirza II 43, 105, 121, 128, 130
Misfortune 81
Miss Agnes 44
Miss Bell 26
Miss Darcy's Grey Wilkes 22
Miss Darcy's Pet Mare 22
Miss Disco 203
Miss Dogwood 178
Miss Euston 26
Miss Hip 22
Miss Matty 11
Miss Pratt 80
Miss Slamerkin (GB) 22
Miss Slammerskin 114
Miss Swapsco 47, 57
Missed the Storm 180
Missy Baba 46
Mister Frisky 109
Mister Slippers 179
Mistress Ford 131
Miswaki 58, 61, 89, 102, 160, 176, 179, 191
Mitochondria & MtDNA 30, 36, 114- 118, 121, 123, 213
Mixed Marriage 46
Mizpah 176
Mogul 11, 49
Moliora 49
Moll 114
Molly Desmond 152
Molly Longlegs 50
Molvedo 108

Monarch 85
Monarque 27
Montagu, Mr. E.W.
Moon Cactus 154
Moonshell 154
Moon Priestess 196
Moonah Mare, The 23
Moore, George 208
Morganette 166
Morocco Barb, The (Old).14, 15, 18, 19, 23, 24, 25, 26, 28
Morris, H.E. 130
Morrissey, John 11
Moslem 79
Mossborough 32, 100
Mostly Sunny 157
Mother Seigel 68, 147, 196
Mother Western 49
Mott, Bill 111
Mountain Cat 43, 61, 101, 183
Mountain Flower 181, 182
Moussac, Paul de 204
Moyglare Stud 204
Mr. Prospector 31, 47, 51, 55, 61, 88, 89, 93, 102, 167, 176, 178, 191, 211
Mr. Right 110
Mrs Ridway 143
Mrs. Butterwick 45, 163
Mrs. Penny 43, 167
Mt. Livermore 58, 93, 98, 169, 181
Mtoto 131, 132
Muhtarram 108
Mujtahid 102, 178
Mukaddamah 101, 205
Mukddaam 58
Muley Moloch 77
Mummy's Pet 95
Mumtaz Begum 13, 43, 51, 67, 121, 128, 170, 178, 197
Mumtaz Mahal 10, 12, 27, 43, 45, 51, 54, 56, 111, 121, 128, 129, 173, 194, 197, 208, 209, 212
Munaaji 180
Munda 130
Muranna Stud 204
Musidora 163
Musket 11
My Babu 46, 57, 107
My Flag 58
My Play 13, 202
My Yellow Bird 43
Myrtlewood 46, 55, 178
Mystery 131
Mystery Rays 191

N

Nadir Shah 16
Nantallah 193
Nashua 58, 88, 98, 178
Nashwan 55, 58, 98, 131, 169, 188, 198
Naskra 98
Nasram 98
Nasrullah 13, 46, 51, 100, 128, 133 144, 167, 168, 170, 173, 176-181, 195, 197 212
Nassipour 169
Natalma 58, 149, 154, 158, 182, 189, 190, 192
Natashka 47
Native Dancer 13, 31, 46, 55, 56, 88,
Native Dancer cont. 96, 102, 154, 178, 179, 182, 189, 203, 213
Native Partner 47, 189
Natural Barb Mare 26
Naturalism 205
Nearco 12, 13, 31, 39, 46, 51-54, 67 80, 85- 88, 96- 101, 112, 114, 121, 133, 145, 151, 156, 166, 169, 173, 176, 189, 195, 197, 201, 208, 216
Nearctic 13, 52, 67, 80, 87, 101, 133, 149, 151, 153, 189, 203
Nebbiolo 98
Neckar 105
Neddie 104
Needles 179
Nera di Bicci 85
Neriad 47, 57, 192
Nerud, John 210
Nervesa 121
Never Bend 47, 57, 98, 154, 173, 179
Never Knock 47, 58
Never Say Die 168
Never Too Late II 193
New Moon 54, 196
Newcastle Turk 15
Newhaven Park Stud 204
Newminster 14, 44, 54, 95, 112, 144,
Newton's Bay Arabian 23
Niarchos, Maria 3, 43, 58, 190, 191, 192, 193
Niarchos, Stavros 136, 150, 190, 191, 192, 193
Niccolo dell'Arca 86, 88, 121, 145, 195, 201
Nichollas, Rick 186
Nijinsky II 13, 47, 52, 56-61, 67, 88-94, 173, 179-183, 188, 191, 210
No Class 47, 58
Noblesse 47, 57
Nodouble 58, 95
Nogara 45, 67, 68, 86, 88, 121, 144, 151, 166
Noholme II 95
Noisette 81
Nokbul 130
Noor Jahan 32, 193
North Ridge Farm 204
Norther Park 43, 101
Northern Baby 90, 101
Northern Dancer 101, 127, 134, 149, 153, 154, 159, 164,173, 176, 179, 180-183, 189-192, 195, 203, 210
Northern Guest 207
Northern Light 200
Northern Sea 182
Northern Taste 47, 181, 182
Northfields 56, 180
Novice 27
Nun's Veil 127
Nureyev 47, 55, 57, 58, 61, 90, 93, 101, 129, 150, 154, 178, 180, 190, 191, 193, 211
Nutswith 79

O

O'Brien, Vincent 57, 184, 186, 209
O'Rourke, Garrett 136
Oak Cliff Th/brds 181
Obliterate 13
Obscurity 114
Octagonal 43, 97, 128, 194, 205
Oh So Sharp 163, 188
Old Bald Peg 14, 16, 18, 19, 22, 24 44, 48, 53, 121
Old Careless 18
Old Child Mare 25
Old Ebony 22
Old Hautboy (Hautboy) 16
Old Merlin 25
Old Morocco Barb 16, 18, 22, 25
Old Morocco Mare 48
Old Polly 25
Old Spot 23, 48
Old Whitefoot 23
Old Whynot 23
Olden Times 103
Oldfield Mare, The 23, 24, 25
Oldham, G. A. 204
Oleander 105
Olymp 106
Omaha 109
Omar Khayyam 11
Omnium II 199
On and On 88
Oncidium 105
Opening Verse 101
Oppenheimer, Sir Philip 204
Orby 13, 60, 83, 147
Orlando 14, 83, 112
Orme 13, 45, 54, 83, 85, 109, 141, 142, 147, 153, 166, 211
Ormonde 13, 90, 88, 109, 142, 162, 163, 166, 167, 202
Ornament 148, 166
Orsenigo 106
Ortello 109
Orton, Peter 136
Orville 80, 81, 166
Osborne, John 18
Our Lassie 171
Overbrook Farm 183
Overbury Stud 204
Owen Tudor 67, 95, 170
Oxford 112
Oxford Dun Arabian 15
Oxygen 80
Oysterfoot Arabian Mare 22
Ozone Friendly 191

P

Paget, Miss Dorothy 204
Pago Pago 96
Palace Music 10, 111, 184, 205
Pampeiro 128
Pancho Villa 73, 88, 100
Panorama 46, 54, 60
Pantaloon 44, 110, 113
Pantheon 13, 104
Papila 46
Papyrus 11, 13, 45, 55, 69, 83, 85, 86, 104, 192, 193, 202
Paradoxical 45
Pardal 96
Pardallo 96
Partner 12, 25, 44, 48, 49, 50, 113
Pas de Nom 89
Pasadoble 191
Pasca 130
Pasch 130
Pass a Glance 191
Past Eight 46
Paulowitz 12, 141

Paulson, Allen 43, 111, 188, 204
Payne-Gallwey, Sir Philip 136, 190
Peaks and Valleys 98, 180
Pearl Cap 165
Pearl Diver 105
Pearl Maiden 45
Peck, Robert 80
Pelham's Barb 15, 17, 23
Penelope 44, 80, 81
Pennekamp 88, 102, 154
Pentire 188
Pepenador 105
Perdita II 166, 167
Perfume II 46
Persian Bold 100
Persian Gulf 88, 105, 131, 132
Persimmon 12, 29, 53, 105, 162, 166
Personal Ensign 58
Personal Flag 58, 87, 109
Personal Hope 101
Peter Pan 104, 145
Peters, Leone J. 204
Peteski 102
Petingo 96
Petite Etoile 163
Petition 46, 54, 86, 96, 198
Phaeton 108
Phalaris 13, 55, 60, 68, 80, 96, 111, 147, 150, 151, 166, 174, 175
Phantom 51
Phaona 175
Phar Lap 14, 76, 207, 209
Pharamond II .13, 32, 54, 96, 121, 172, 174, 178, 189, 196
Pharis II 13, 54, 96, 174, 190, 199
Pharly 169
Pharos 13, 45, 54, 55, 67, 68, 80, 88, 151, 173, 195, 197
Phideas 150
Phillips, John 183
Phipps Ogden 45, 204
Phipps Ogden Mills 204
Phone Trick 61, 101, 157
Piersfield Mare 27
Piggot, Lester 208
Pilate 13, 104
Pilgrimage 68, 69, 133, 150, 170, 175, 196
Pine Bluff 31, 61, 89, 101
Pink Domino 45
Piping Peg 24
Piping Rock 171
Pirate Army 97
Pistolet Blue 100
Pitcairn 66
Place's White Turk 14, 15, 16, 17, 18
Place's White Turk 21, 22, 24, 44
Place, Mr. 14. 17
Plagiarizing 157
Plaisanterie 83, 85, 150, 153
Plaudit 104
Playlist 179
Playmate 176
Pleasant Colony 108, 133, 134, 164
Pleasant Stage 123
Pleasant Tap 58, 108, 133
Plenipotentiary 77
Plesch, Madame Arpad 204
Pliohiippus 35
Plucky Liege 32, 45, 52, 54, 73, 121, 129, 130, 151, 166
Pocahontas 11, 12, 31, 40, 44, 50, 53, 56, 76- 85, 127, 143, 144, 146, 150-155, 166, 167, 176, 183, 197, 202, 203
Poetess 85
Polish Navy ..89, 101
Polish Patriot 89
Polish Precedent 89, 101
Polymelian 55
Polymelus 13, 29, 55, 68, 80, 96, 147, 148, 171, 174, 175, 188
Polynesian 13, 46, 55, 88, 96, 102
Pommern 148
Ponte Vecchio 181
Porter, Alan 86, 139
Posse 190
Pot 8 O's 25, 81, 112
Practicante 86, 107
Prather, John 139
Pratt's Mare 49
Precipitation 46, 67, 103
Premiership 102
Presenting 131
Prestonwood Farm 204
Pretense 96, 178
Pretty Lady 150
Pretty Polly 23, 152, 164
Priam (by Emilius) 13, 142
Priam II 51, 83
Priamos 106
Pricket 115, 188
Primero 129
Primo Dominie 100
Prince Batthyany 52
Prince Bio 108, 154
Prince Charlie 167
Prince Chevalier 12, 153
Prince Chimay 108
Prince John 108
Prince Palatine 108, 197
Prince Rose 12, 86, 176
Princely Gift 98
Princequillo 12, 46, 55, 56, 57, 72, 75, 83, 86, 108, 112, 148, 176, 179, 181, 192, 193, 197, 198
Princess Catherine 85
Prior, C.M. 19
Prior, Miss C.B. 26
Priscilla 175
Private Account 87, 109, 179
Private Terms 87, 109, 184
Prized 97
Promisied Land 181
Promising (NZ) 149
Pronto 107
Prospector's Gamble 102
Proud Clarion 164
Proud Truth 108
Prove Out 191
Prunella 44, 50, 53, 81
Prunus 106
Pulleine's Chestnut Arab 15, 24, 26
Purity 79
Pursuit of Love 180
Pussy 80
Q
Qirmazi 191
Quack 181
Queen's Hussar 96
Queen's Statute 46
Queen Anne;s Moonah Barb 23
Queen Elizabeth II, H.R.H. 58
Queen Mab 166
Queen Mary 44, 54, 144, 175
Queena 89
Quest for Fame 98, 163, 180, 205
Questionnaire 104
Quicklime 79
Quicksilver Mare 27
Quiet American 32, 102, 149, 186
Quiver 45, 175
R
Rabelais 12, 45, 52, 55, 68, 108, 112, 151, 159, 169, 211
Radium 14
Raeburn 151
Rahy 57, 98, 149, 169, 180, 183, 188
Rainbow Quest 47, 57, 58, 98, 149, 169, 180
Raise a Man 102
Raise a Native 58, 88, 93. 102, 178
Raise the Standard 190
Raise You 178
Raised Socially 102
Ramsinga 157
Rare Mint 183, 191
Rassmussen, Leon 41
Rataplan 13, 31, 44, 50, 76, 79, 80, 144, 145, 146, 152, 161, 167
Ravinella 89
Rayon d'Or 79, 167
Razyana 149, 212
Red Carnival 43
Red God 88, 98, 168, 169, 170, 189
Red Ransom 88, 97
Red Ray 173
Redtop 107
Reel 44
Reference Point 131
Reflected Glory 96
Regal Classic 58
Regret 164
Regulus 11, 24, 44, 48, 49
Reigh Count 110, 145
Reine Isaure 130
Relaunch 61, 93, 103
Reliance II 67, 200
Relic 103, 131, 132
Relko 67, 200
Requested 104
Restless Native 102
Retreat 152
Reve d'Or 163
Revidere 181
Rhoda B. 147
Rhythm 89
Rialto 170
Ribot 12, 39, 46, 56, 57, 85, 93, 108, 111, 112, 133, 149, 156, 158, 165, 167, 192, 193, 201, 208
Richardson, Chris 136
Richardson, Tim 136, 190, 193
Riddlesworth 77
Riding Rays 173
Right of Way 86
Rightaway 152, 153
Risen Star 88, 100
Riva Ridge 179
River Prince 25

River Spcial 61
Riverman 47, 57, 86, 98, 159, 178, 179, 181
Riverside 47
Roanoke 108, 134
Robellino 97
Robert Le Diable 142
Roberto 47, 57, 88, 93, 97
Robin des Pins 101, 183, 191
Rock City 191
Rock Sand 13, 45, 83, 85, 104, 111, 112, 132, 146, 153, 162, 264, 175
Rockefealla 130
Rockefel 130, 163, 166
Rogers, Capt. D. 204
Rogers, Jack 128
Roi Herode 63, 110
Roman 109, 212
Roman, Dr, Steven 41
Romanella 111
Rondeau 87
Roquebrune 85, 176
Rory's Jester 205
Rosaline 68
Rosalys 68
Rosananti 169
Roscicrucian 144, 150, 196
Rose Bower 181
Rose Prince 108
Rose Red 132
Rose Vert 130
Rosebery 110
Rosebery, Earl of 192, 204
Roseden 26
Rosedrop 68, 69, 196
Rosenberg, Dan 136
Roseville 104
Rosy Cheeks 79
Rothschild, Anthony de 129
Rothschild, Baron Edouard de 130
Rothschild, Baron Guy de 204
Rouge Rose 44, 144
Rough'n Tumble 104
Rough Shod II 46, 55, 192, 211
Round Table 55,108
Rout, Cuthbert 19
Rowdy Angel 157
Rowdy Angel 61
Roy 103
Royal Charger 12, 13, 43, 46, 51, 54,
Royal Mare, The 23, 24, 25
Royal Natural Barb Mare 25
Royal Palm 130
Royal Serenade 130
Rubiano 102
Ruby Mare, The 49
Ruffian 208
Runaway Bride 168, 170, 189
Runaway Groom 58, 98, 169, 180
Runnymede Farm 204
Rupert 17
Russell, Dr. Peter J. 116, 121
Rustom Pasha 106
Rutland's Barb 14
Ryan, Kate 189
Ryan, Mike 136

S

Sadler's Wells 47, 55, 58, 90, 101, 129, 191
Sainfoin 13, 31, 45, 54, 67, 68, 85
Sainfoin cont. 104, 111 , 145, 150, 153, 162, 166, 175, 193
Saint Cyrien 107, 149
Salmon Trout 12, 110
Saltram 50, 81, 141
Saluda 68
Samos 130
San San 165
Sanctimony 79
Sanctus 32, 172
Sanda 85, 111, 141, 145, 146, 166
Sandal 44, 54, 111, 145
Sanelta 192
Sangster, R. 57, 184, 186, 204, 205
Sans Souci II 79, 130, 154, 155
Sansonnet 54, 67, 170, 198
Sansovino 67, 128, 150, 172, 195
Santa Cruz 196
Sarazen 145
Sarita 151, 152
Saros 103
Sassafras 1-3
Satirical 68
Saumarez 98, 165
Saumur 85
Sayajirao 13, 67, 79, 115
Sayyedati 192
Scamp 127
Scapa Flow 68, 195
Sceptre 163
Schossberg 104
Scott, Alex 188
Scottish Union 130, 164
Sea Bird II 88, 102, 165, 182
Sea Hawk II 106
Sea Hero 89, 101, 165, 199
Sea Saga 182
Sea Sands 43
Search for Gold 89
Seaside Attraction 43
Seattle Slew 55, 93, 100, 111, 164
Seattle Song 100
Secretariat 56, 64, 72, 73, 75, 86, 88, 100, 164, 165, 167, 179, 181, 209
Secreto 90, 101, 163
Sedbury Royal Mare 22
Sedbury Turk 14, 15 18
Seeking the Gold 58, 89, 102, 160, 176, 179
Segenhoe Stud 204, 205
Selaby Turk 15, 21, 24
Selene 32, 45, 54, 58, 67, 69, 73, 121, 128, 144, 151, 159, 166, 167, 170, 189, 193, 195, 196
Selim 12, 44, 80, 81, 113
Selima 11, 26, 49, 50
Selkirk 102
Septieme Ciel 100, 191
Sequence 178
Serena's Song 183
Serenissima 32, 68, 128
Seymour, Lord Henry 27
Shaamit 105, 131, 135, 163
Shadeed 101, 180, 186, 188
Shaftersbury Turk 14, 21, 24
Shahrastani 127, 163
Shalford 61
Sham 75
Shama 182
Shamikh 180
Shareef Dancer 188
Sharpen Up 47, 88, 102
Sharpo 102
Sheikh Albadou 101, 186
Shernazar 105
Sheshoon 103
Shield's Galloway 19, 20, 26
Shikampur 108St. Jovite 108
Shirley Heights 98
Shoemaker, Bill 208
Siberia 79
Sibola 68, 69, 201
Sicambre 46, 108, 112
Sicarelle 163
Sickle 13, 32, 54, 96, 121, 172, 174, 178, 196
Sierra 13, 45, 111, 145, 150, 151, 176, 193, 197
Signal Light 174
Signe Divine 154
Sillery 169, 180
Silver Buck 26
Silver Fog 26
Silver Hawk 88, 97
Silver Screen 108
Silver Supreme 192
Silvio 80
Simona 68
Sing Sing 95
Sir Archy 12, 27
Sir C. Turner's White Turk 15
Sir Gallahad III 13, 32, 45, 54, 57, 74, 83, 86, 87, 109, 121, 142, 145, 151, 176, 178, 179, 193, 211
Sir Gaylord 56, 62, 86, 88, 97, 178
Sir Hales Bay Turk 23
Sir Hercules 80, 83, 112
Sir Ivor 58, 97, 179
Sir John Sebright's Arabian mare 26
Sir Marmaduke Wyville's Grey 19
Sir Paul 141
Sir Peter 12, 44, 81, 166
Sir Tristram 31, 47, 54, 58, 88, 97, 196, 205
Sir Victor's Barb 14, 19
Sir Willia's Turk 15
Sir William Morgan's Black Barb 14
Sir Winborne 97
Sissian, Michael 205
Sister Sarah 67, 151, 152, 154, 203
Skinner, J. S. 11
Skip Trial 87, 109
Sky Beauty 169
Sky Classic 58, 101, 127, 180, 181
Slew O'Gold 73, 100
Slide Rule 104
Slip Anchor 98
Smarten 97
Smile 103
Smirke, Charlie 129
Smokey Eyes 67
Snailwell Stud 204
Snake 49
Snap 22, 50
Snark 104
Snorting Bess 21
Snow Bride 169, 188
Snurge 164
Solar Slew 111, 182
Solario 14, 51, 79, 129, 150, 151
Solon 11, 24, 103, 113
Somerset mare 22

Somethingroyal 46, 56, 72, 86
Son-in-Law 14, 67, 106, 112
Song 95
Sonic Lady 188
Sorcerer 113, 142, 166
Sostenuto 149
Soul Mate 168
Soul of the Matter 184
Source Sucree 55, 173
Sovereign Dancer 90, 101
Sovereign Path 98
Spain Lane 179
Spalding, Bobby 136
Spanker 18, 24, 36, 44, 48, 49, 50
Spanker Mare, The 24
Spearmint 12, 32, 68, 69, 88, 147, 162, 171
Special 190, 192
Spectacular Bid 64, 88, 100
Spectator 50
Speculum 12, 110, 112
Spend a Buck 86, 96, 165
Spendthrift 26, 103, 113
Spike Island 128
Spiletta 11, 22, 49
Spinning World 193
Spion Kop 147, 162
Spring Adieu 149, 170
Spring Run 189
Springfield 85, 111, 141, 146, 166
Spur 104
Spy Song 145
Squire Jones 179
Squirt 44, 49, 112
St. Angela 52, 68, 83, 196
St. Damien 148
St. Denis 148
St. Frusquin 12, 52, 68, 69
St. Germans 105
St. Honorat 153
St. Josephine 148
St. Jovite 133
St. Marguerite 85
St. Martin 23
St. Paddy 95, 162, 164
St. Serf 12, 155
St. Simon 12, 22, 29- 32, 40, 45, 52, 54, 55, 67- 69, 79, 83, 85, 87, 108, 112, 142-156, 166, 174, 175, 195, 196, 201, 203, 211, 216
St. Thomas 25
St.Albans 13, 79
Stalwart 108
Star Appeal 106
Star de Naskra 98
Star Kingdom 46, 58, 61, 67, 95, 111, 135, 159, 166, 181, 205
Star Way 106
Stardust 61, 67, 95
Steel, Dr. J. D. 71, 76
Stefan the Great 12, 110
Stellarette 43
Sterling 112
Stewart, Dr. Anthony 71, 76
Stockwell 13, 31, 40, 44, 50, 53, 76,
Stop the Music 97
Storm Bird 13, 47, 57, 58, 61, 90,
Storm Cat 43, 61, 72, 73, 92, 93, 101, 158, 180, 183
Straight Deal 199
Strawberry Road 101
Strawbridge, George Jr. 204
Strike the Gold 88, 102
Striking 46, 199
Stripling 23
Strodes Creek 97
Stroud, Anthony 186
Stymie 14
Suave Dancer 165
Successor 181
Sueborg 200
Suicide 141
Sultan (Arabian) 16, 18, 22, 44
Sultan (by Selim) 80- 83, 113, 146
Sultry Song 88, 97
Summer Squall 61, 101
Summertime 46, 103
Sun Again 87, 109, 203
Sun Chariot 198
Sun Colony 133
Sun Princess 43, 51, 128
Sun Reigh 12, 110
Sun Teddy 13, 83, 87, 109, 142
Sunday Silence 47, 55, 58, 90, 97, 153, 165, 181
Sundridge 12, 13, 43, 45, 60, 110, 111, 145, 151, 193
Sunglow 88, 109
Sunny's Halo 88, 97, 156, 193
Sunrise Flight 133
Sunshot 45
Sunstar 12, 45, 110, 111
Surumu 106
Sutton Turk 15
Swain 188
Swaps 26, 27, 46, 56, 67, 88, 95, 164, 171, 178, 208
Sweep 45, 104
Sweep On 26
Sweet Tooth 88
Swing and Sway 104
Sword Dancer 26, 57, 69, 87, 88, 109
Swynford 13, 54, 67, 87, 105, 111, 127, 128, 132, 133, 150-152, 170, 178, 195, 196
Sykes, Sir Mark 153

T

T.V.Lark 98
Tabaris 129
Tabasco Cat 101
Tabriz 128
Taffolet Barb or Bustler 14, 15, 18, 19, 24, 25
Talinum 88, 102
Tamerett 47, 57
Tamerlane 105
Tan Pronto 107
Tanerko 87, 109, 200
Tantieme 67, 87, 109, 165, 166, 200
Tartar 19, 20, 26, 49, 50, 113
Tasso 73
Taylor, Duncan 136
Taylor, Edward P. 203, 210
Technology 87, 109, 211
Teddy 12, 13, 45, 54, 57, 83, 85, 86, 87, 109, 151, 176, 178, 199, 202
Tehran 108
Tel Quel 105
Teleferique 110
Temperence Hill 97
Tenerani 108
Tennyson 105
Tentam 57, 103
Teresina 13, 67, 85, 193
Terimon 105
Terlingua 31, 72, 73, 193
Termagent 81
Ternynck, Jean 204
Tersa 89
Tesio, Federico 31, 39, 68, 85, 114,
Tessa Gillian 43, 46, 54, 128, 212
Tetratema 12, 110, 171
Thatch 61, 95
Thatching 95
The Axe II 105, 134
The Baron 53, 77, 79, 112, 144, 161
The Boss 13, 60
The Colonel 79
The Duke 79
The Flying Dutchman 12, 77, 113, 143
The Minstrel 55, 57, 90, 101, 111, 160, 169
The Nun 27
The Palmer 144, 150, 196
The Panther 85
The Peer, 11
The Recorder 54
The Rhymer 105
The Sailor Prince 68
The Scout II 151
The Slave 144
The Squaw II 46
The Tetrarch 12, 43, 45, 54, 60, 63
The Wicked North 101
Theatrical 93, 101
Theft 129
Thong 192, 193
Thormanby 110, 113, 144, 155
Three Troikas 167
Thunder Gulch 102, 165
Tiajuana 73
Tiberius 111, 129, 130
Ticino 105
Tide-way 196
Tight Spot 133
Tim Tam 164, 167, 179
Timber Country 178
Time Charter 163
Time for a Change 87, 109
Time to Explode 101
Timeless Moment 87, 109
Timor 107
Tinner's Way 88
TOBA 138
Tom Fool 13, 46, 56, 96, 178, 189
Tom Rolfe 57, 108
Tomato 79
Tony Bin 131
Too Chic 169
Top Ville 100
Topanhoora 107
Topiary 83, 85, 153
Topsider 90
Touchstone 14, 23, 29, 83, 95, 112, 144, 166
Tough as Nails 158
Toulon 164
Tourbillon 12, 27, 45, 54, 55, 67, 83, 85, 107, 190, 198
Tout Suite 67
Tracer 13

Tracery 13, 31, 45, 58, 67, 69, 83, 85, 86, 104, 111, 128, 152, 154, 181, 182, 183, 192, 193, 201, 202, 212
Traffic 13, 85
Traffic Cop 95
Traffic Judge 46, 67, 95
Tramp 50, 77, 80, 81
Trampoline 77, 80, 81, 83, 146
Transvaal 85
Trap Rock 13, 104
Traverse 13, 85
Treasure 21
Tregonwell's Nat.Barb Mare 16, 21 24, 121
Trempolino 102
Trentham 146
Trenton 68
Tri-Jet 96
Tricky Creek 101
Trigo 127, 128
Triptych 131
Tristan 127, 148, 192
Troutbeck 164
Trumpator 80, 81, 113
Try My Best 101
Tudor Melody 95
Tudor Minstrel 46, 54, 67, 95, 145, 149, 170, 173, 189, 198, 208
Tulloch (NZ) 64, 209
Tulyar 40, 64
Turkoman 88, 102
Turn-To 46, 54, 57, 88, 90 97
Twaite's Dun Mare 25
Twenty Grand 164
Twig Moss 107
Twilight Agenda 88, 97
Twilight Tear 203
Twining 102
Two True Blues, The 21

U

Udaipur 128
Uganda 45
Ulbrich, Richard 139
Ultimus 145
Umatilla 179
Umidwar 46, 129, 150, 170, 171
Unbreakable 13, 55, 88, 96
Unbridled's Song 61
Unbridled 51, 61, 102, 165, 183, 194
Understanding 181, 182
Unfuwain 55, 58, 90, 101, 188
Upas 83, 85
Urban Sea 179
Utopico 107

V

Vaguely Noble 95
Vahren 63
Vain 96
Val de Loir 105, 179
Valiant Nature 108
Valid Appeal 61, 103
Valley Crossing 109
Vampire 26, 85, 142
Vandale II 106
Vanderbilt, Alfred G. 203
Varola, Dr. Franco 41
Vatellor 12, 108
Vatout 12, 108, 130
Vedette 12, 23, 108, 110, 110, 142, 143, 144
Velocipede 142, 143
Vertex 105
Vettori 61, 102
Vice Regent 93, 101, 179
Victoria Park 154, 178
Vieux Manoir 105
Vigilant 85
Vimy 131, 132
Vincent, John 136
Vintner Mare, The 121
Vintner Mare, The 16, 22, 25, 48
Virgin 22
Virginia Water 173
Viridiflora 130
Viridis 146
Vispania 111
Vitamine 130
Vixen 25
Volley 144
Voltaire 12, 44, 53, 112, 143, 144
Volterra, Madame L. 203, 204
Voltiguer 12, 112, 143, 162
Vuillier, Colonel 40, 41, 115
Vuillier, Madame 41, 115, 172
Vulture 83

W

Waffles 147
Waldenstein 106
Walmac Internat. Farm 190
Walter Willy 191
Walton 12
Waltz 80
Walwyn, Peter 190
Waquoit 103
War Admiral, 11, 108,164, 178, 179
War Relic, 11, 61, 103, 113
Warlock's Galloway 19
Warning 57, 61, 103, 157
Water Bank 98
Wavering Monarch 61, 102, 157
Waxy 44, 80, 81, 112, 144
Web 80, 81, 144
Wellingtonia 31, 79, 83, 85, 141, 150, 152, 153, 154, 202, 203
Welsh Pageant 95
Wenlock 111, 145
Wentworth, Lady 3, 16-22, 24, 29
Wertheimer, J. 204
West Australian, 11, 103, 113
Whakalyric 191
Whalebone 80, 112, 144
Wharton Mare, The 23
What a Pleasure 100
Whirlaway 128, 164, 203
Whisk Broom II 104
Whisker 80, 146
Whiskey 44, 50, 51, 80, 81
White Eagle 127
White Gloves 179
White Legged Lowther Barb 14, 48
White Lodge Stud 204
Whitefoot 22, 26
Whitelock 112
Whitenose 11
Whitenose Mare 26
Whiteshirt 22
Whitney, Cornelius Vanderbilt 202
Whitney, Harry Payne 202
Whitney, John Hay 202
Whizgig 80
Widden Stud, 204
Widener, Joseph E. 203
Wild Again 98, 101
Wild Harmony 101
Wild Risk 46, 54, 55, 58, 61, 108, 131, 132, 170, 183, 188, 189, 210
Wild Violet 131, 170
Wildair 114
Wild Dayrell 12
Wildenstein, Alec, 139, 204
Wildenstein, Daniel 43, 149, 204
Wilkes (Fr) 96
Will Somers 95
William's Arabian 15
William of Valence 108
William The Third 45, 147
Williams, Professor 167
Williams, Roger 11
Williamson's Ditto 81
Williamstown 100
Windfields 203
Windhound 110, 113, 144
Windsor Lad 64, 67, 105, 127, 128,
Windsor Slipper 105, 129
Winning Colors 165, 207
Wire 80
Wisdom 79, 142, 152, 153
Wishing Well 181, 182
With Approval 98
Woful 80
Wolfhound 61, 101, 157, 188
Wood Violet 131, 170
Woodbine 44, 54, 144, 155
Woodlands Stud (Aust) 204, 205
Woodman 58, 61, 89, 102, 160, 176,
Woodpark Stud 189
Woodpecker 12, 81, 110 113
Woods of Windsor 178
Woodward, William 151, 203
Woolums, M "Sissy" 138
Worden II 108
World Appeal 103
Wright, Warren 203
Wychwood Abbot 26
Wyndham 128
Wynn Arabian 15

XY

Yellow God 98
Yellow Turk 18, 24
Yoshida T. 186
Yoshida, Zenya 204
You'd be Surprised 169
Young Cade 26
Young Generation 95
Young Giantess 50, 81
Young Noisette 81
Young True Blue 22
Young, William T. 43, 92, 183, 192, 194, 195, 204

Z

Zabara 132
Zabeel 88, 97
Zafonic 61, 102
Zamazaan 105
Zank 105
Zariba 190
Zeddaan 98, 168
Zev 164
Zilzal 101, 188
Zivato 168